PLANNING ETHICALLY RESPONSIBLE RESEARCH

Applied Social Research Methods Series
Volume 31

APPLIED SOCIAL RESEARCH METHODS SERIES

1. **SURVEY RESEARCH METHODS**
   Second Edition
   by FLOYD J. FOWLER, Jr.
2. **INTEGRATING RESEARCH**
   Second Edition
   by HARRIS M. COOPER
3. **METHODS FOR POLICY RESEARCH**
   by ANN MAJCHRZAK
4. **SECONDARY RESEARCH**
   Second Edition
   by DAVID W. STEWART
   and MICHAEL A. KAMINS
5. **CASE STUDY RESEARCH**
   Second Edition
   by ROBERT K. YIN
6. **META-ANALYTIC PROCEDURES FOR SOCIAL RESEARCH**
   Revised Edition
   by ROBERT ROSENTHAL
7. **TELEPHONE SURVEY METHODS**
   Second Edition
   by PAUL J. LAVRAKAS
8. **DIAGNOSING ORGANIZATIONS**
   Second Edition
   by MICHAEL I. HARRISON
9. **GROUP TECHNIQUES FOR IDEA BUILDING**
   Second Edition
   by CARL M. MOORE
10. **NEED ANALYSIS**
    by JACK McKILLIP
11. **LINKING AUDITING AND METAEVALUATION**
    by THOMAS A. SCHWANDT
    and EDWARD S. HALPERN
12. **ETHICS AND VALUES IN APPLIED SOCIAL RESEARCH**
    by ALLAN J. KIMMEL
13. **ON TIME AND METHOD**
    by JANICE R. KELLY
    and JOSEPH E. McGRATH
14. **RESEARCH IN HEALTH CARE SETTINGS**
    by KATHLEEN E. GRADY
    and BARBARA STRUDLER WALLSTON
15. **PARTICIPANT OBSERVATION**
    by DANNY L. JORGENSEN
16. **INTERPRETIVE INTERACTIONISM**
    by NORMAN K. DENZIN
17. **ETHNOGRAPHY**
    by DAVID M. FETTERMAN
18. **STANDARDIZED SURVEY INTERVIEWING**
    by FLOYD J. FOWLER, Jr.
    and THOMAS W. MANGIONE
19. **PRODUCTIVITY MEASUREMENT**
    by ROBERT O. BRINKERHOFF
    and DENNIS E. DRESSLER
20. **FOCUS GROUPS**
    by DAVID W. STEWART
    and PREM N. SHAMDASANI
21. **PRACTICAL SAMPLING**
    by GARY T. HENRY
22. **DECISION RESEARCH**
    by JOHN S. CARROLL
    and ERIC J. JOHNSON
23. **RESEARCH WITH HISPANIC POPULATIONS**
    by GERARDO MARIN
    and BARBARA VANOSS MARIN
24. **INTERNAL EVALUATION**
    by ARNOLD J. LOVE
25. **COMPUTER SIMULATION APPLICATIONS**
    by MARCIA LYNN WHICKER
    and LEE SIGELMAN
26. **SCALE DEVELOPMENT**
    by ROBERT F. DeVELLIS
27. **STUDYING FAMILIES**
    by ANNE P. COPELAND
    and KATHLEEN M. WHITE
28. **EVENT HISTORY ANALYSIS**
    by KAZUO YAMAGUCHI
29. **RESEARCH IN EDUCATIONAL SETTINGS**
    by GEOFFREY MARUYAMA
    and STANLEY DENO
30. **RESEARCHING PERSONS WITH MENTAL ILLNESS**
    by ROSALIND J. DWORKIN
31. **PLANNING ETHICALLY RESPONSIBLE RESEARCH**
    by JOAN E. SIEBER
32. **APPLIED RESEARCH DESIGN**
    by TERRY E. HEDRICK,
    LEONARD BICKMAN,
    and DEBRA J. ROG
33. **DOING URBAN RESEARCH**
    by GREGORY D. ANDRANOVICH
    and GERRY RIPOSA
34. **APPLICATIONS OF CASE STUDY RESEARCH**
    by ROBERT K. YIN
35. **INTRODUCTION TO FACET THEORY**
    by SAMUEL SHYE
    and DOV ELIZUR
    with MICHAEL HOFFMAN
36. **GRAPHING DATA**
    by GARY T. HENRY
37. **RESEARCH METHODS IN SPECIAL EDUCATION**
    by DONNA M. MERTENS
    and JOHN A. McLAUGHLIN
38. **IMPROVING SURVEY QUESTIONS**
    by FLOYD J. FOWLER, Jr.
39. **DATA COLLECTION AND MANAGEMENT**
    by MAGDA STOUTHAMER-LOEBER
    and WELMOET BOK VAN KAMMEN
40. **MAIL SURVEYS**
    by THOMAS W. MANGIONE

PLANNING ETHICALLY RESPONSIBLE RESEARCH

A Guide for Students and Internal Review Boards

Joan E. Sieber

Applied Social Research Methods Series
Volume 31

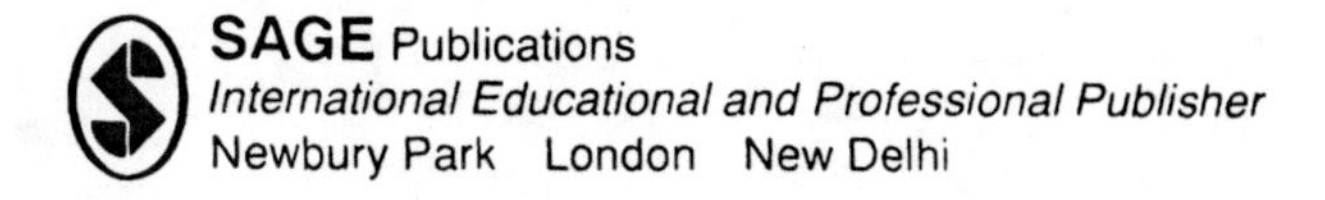
SAGE Publications
International Educational and Professional Publisher
Newbury Park London New Delhi

For information address:

SAGE Publications, Inc.
2455 Teller Road
Newbury Park, California 91320
E-mail: order@sagepub.com

SAGE Publications Ltd.
6 Bonhill Street
London EC2A 4PU
United Kingdom

SAGE Publications India Pvt. Ltd.
M-32 Market
Greater Kailash I
New Delhi 110 048 India

Printed in the United States of America

Library of Congress Cataloging-in-Publication Data

Sieber, Joan E.
Planning ethically responsible research: a guide for students and internal review boards/Joan E. Sieber.
p. cm. — (Applied social research methods series : v. 31)
Includes bibliographical references and index.
ISBN 0-8039-3963-9 (cl.) — ISBN 0-8039-3964-7 (pb.)
1. Research—Moral and ethical aspects—Handbooks, manuals, etc.
I. Title. II. Series.
Q180.55.M67S54 1992
174' .93—dc20 92-15340
CIP

00 01 02 03 11 10 9 8 7

Sage Production Editor: Judith L. Hunter

Contents

To the memory of my mentor,
John T. Lanzetta

Preface

This book was written to provide social scientists, their students, and members of research ethics committees with the theory and practical knowledge needed to plan ethically responsible social and behavioral research. It interprets current viewpoints on what ethical research is, especially those views presented in *The Belmont Report* (National Commission, 1978), a document set forth by the National Commission on the Protection of Human Subjects of Biomedical and Behavioral Research. It is also a practical handbook on how to translate ethical principles into valid research methods and procedures that satisfy both scientific and ethical standards. Parts of each chapter provide guidelines for satisfying federal regulations governing human research and for working with one's Institutional Review Board (IRB), or Human Subjects Committee, as such groups are variously called.

Federal regulations of human research require IRB review of research plans (protocols), but are sparse on the matter of how to meet IRB requirements. As IRB chair at California State University, Hayward (CSUH), I spent much time explaining to exasperated investigators how to revise their protocols. Because it is preferable for investigators to know what is required of them in the first place, I decided to write this book.

This book's unpublished predecessor, *The IRB Handbook*, was written, with partial support from CSUH, for CSUH investigators. A grant from the National Science Foundation (BBS-8911646) provided the resources for a needs assessment, consultation with social scientists and their students on the adequacy of an earlier draft of this book, and some released time for writing. To NSF, and to Rachelle Hollander and Vivian Weil, who have monitored this project, I express my gratitude. The needs assessment survey of social science IRBs indicated that chapters on research on children and on AIDS should be added. A grant from the Universitywide Task Force on AIDS, U. C. Berkeley (C89CSH01) provided funding for a conference that examined current solutions to problems of community-based social research related to AIDS.

I thank the many social scientists, IRB members, and students who have critiqued this book in its earlier stages. In particular, statistician Bruce Trumbo and psychologist Mary diSibio provided superb advice and editing throughout the manuscript preparation process. Others, each of whom provided reams of useful criticism of the penultimate manuscript, included Jeffrey Cohen, Ross Conner, Jane Croley, Mary deChesney, Jane Close Conoley, Joni Grey, Maureen Hester, Erica Heath, Carol Jablonski, Sue Hoppe, Paul Kakagawa, Gerald Koocher, Suzanne Kusserow, Paula Knudson, Gary B. Melton, Joan Porter, Pat Schwirian, Ada Sue Selwitz, Marlene Wagner, and Tammy Wall. Working with Sage series editor Debra Rog and senior editor C. Deborah Laughton has been a delight. To all who helped, I express my deep gratitude.

PART I

Research Ethics, Regulations, and the IRB

Research is a complicated activity in which it is easy for well-meaning investigators to overlook the interests of research participants—to the detriment of the participants, scientists, science, and society. To provide a broader ethical and legal perspective on research on human subjects, the federal government, with the help of scientists, philosophers, and lawyers, considered what ethical principles should govern human research. Federal regulations of human research were promulgated and Institutional Review Boards (IRBs) were mandated to review proposed research and ensure compliance with the regulations. IRBs are to be established at institutions where any federally funded research is conducted, except that an institution can, with federal approval and an interinstitutional agreement, use another institution's IRB.

Building on the ethical principles set forth, this book offers a logical ethical framework to guide investigators and those responsible for the ethical review of research. It shows how the methods and procedures of social science can be tailored to meet ethical and legal requirements.

Chapter 1 discusses why there is a broad-based concern for ethics in social and behavioral research, and why unethical social science is bad social science. It explains how and why IRBs were established and how the individual investigator relates to the IRB.

Chapter 2 introduces the "IRB protocol," that is, the written plan of action through which the investigator describes the research to the IRB, addresses its ethical considerations, and indicates what steps will be taken to comply with legal and ethical requirements. The elements of the IRB protocol are briefly presented, and the reader is directed to subsequent chapters of this book that deal with each element. Chapter 12 provides detailed instructions on development of an effective protocol.

Chapter 3 introduces the ethical principles of human research set forth by the National Commission for the Protection of Human Subjects of Biomedical and Behavioral Research (1978) in *The Belmont Report*. One need not be very astute to notice that these principles instruct one to "do it ethically," but they do not say how. The rest of this book is about how: Part II discusses methods of handling issues of consent, privacy, confidentiality, and deception. Part III shows how to assess risk and benefit, and optimize research outcomes. Part IV focuses on ways to respect the needs and interests of two particularly vulnerable research populations, children and the urban poor, with special attention to those at risk for HIV infection. Part V describes how to summarize pertinent ethical considerations in a research protocol.

1

Research Ethics and IRBs

1.1 ETHICS IN SOCIAL AND BEHAVIORAL RESEARCH

Ethics (from the Greek *ethos*, "character") is the systematic study of value concepts—"good," "bad," "right," "wrong"—and the general principles that justify applying these concepts.

Thus, the ethics of social research[1] is not about etiquette; nor is it about considering the poor hapless subject[2] at the expense of science or society. Rather, we study ethics to learn how to make social research "work" for all concerned. The ethical researcher creates a mutually respectful, win-win relationship with the research population; this is a relationship in which subjects are pleased to participate candidly, and the community at large regards the conclusions as constructive. Public policy implications of the research are presented in such a way that public sensibilities are unlikely to be offended and backlash is unlikely to occur.

In contrast, an ethically insensitive researcher may leave the research setting in pandemonium. The ensuing turmoil may harm the researcher, his or her institution, and even the cause he or she seeks to promote, as suggested by the following fictionalized case, adapted from an actual study.[3]

> *Case 1.1: Working at Cross-Purposes.* A researcher sends bilingual research assistants to interview poor Chicano families in Texas about their attitudes toward their children's school. The purpose of her research is to gather information that will help local schools meet the needs of children from families that have recently moved there. Unbeknown to the researcher, many of those interviewed are illegal aliens, who suspect that the research is connected with the U. S. Immigration authorities. They fabricate many of their answers to hide their illegal status here, and they are especially careful to reveal nothing about their children's needs or problems with the school.

Others in that community carefully avoid the researcher, thus ruining the random sampling design.

A better scientist would have understood that community-based research cannot be planned or conducted unilaterally, and that culturally sensitive approaches are required. He or she would have enlisted community leaders in formulating the research procedures, trained appropriate members of the community to assist with conducting the interviews, and closely supervised the entire process, as exemplified in Case 1, Part III (page 77) and discussed in Chapter 11.

Research designs and procedures that result in failure to treat subjects with respect are likely to yield misleading, inconclusive, or biased results. Yet, enlightened self-interest does not come easily to social scientists because they have been trained, typically, to focus on their predetermined research agenda and to ignore the perceptions and expectations of their subjects and of society at large. This "get data" mentality often produces invalid data.

Selective perception plays an important role in the judgment of scientists who are involved in the intense and demanding enterprise of research. When a researcher is narrowly focused on completing a research project, it is easy to overlook some of the interests and perspectives of the research participants and of society at large. In settings where social scientists have unilateral power to conduct research, they may *appear* to get away with insensitivity to the perceptions and expectations of their subjects, but they do not. Insensitive researchers themselves become an integral part of the stimulus array; thus, it should come as no surprise that their subjects often respond with lies and subterfuge. Clearly, sound ethics and sound methodology go hand in hand.

Scientists, themselves, (e.g., Kelman, 1968; Vinacke, 1954) have critically examined some of the ethically questionable assumptions and practices of social research and recommended changes, but it was the federal government that finally brought these issues most forcibly to our attention. In 1974 the federal government mandated the establishment of Institutional Review Boards (IRBs) at all universities that accept funding from the Department of Health and Human Services (DHHS).[4] The role of the IRB is to examine all proposals for research involving human subjects to determine whether the rights and welfare of the subjects are adequately protected. Before starting research, the investigator submits a protocol to the IRB. The protocol describes the proposed research and the arrangements that have been made to ensure that the project adheres to sound ethical and scientific principles.

The wise researcher uses the protocol as a guide for improving the research design and procedures. Chapter 2 briefly introduces the elements of a typical research protocol and directs attention to those parts of this book that will guide the researcher through each part of the protocol.

1.2 WHAT IS AN IRB?

An IRB, or Human Subjects Committee, is a committee mandated by the National Research Act, Public Law 93-348, to be established within each university or other organization that conducts biomedical or behavioral research involving human subjects and receives federal funding for research involving human subjects. The purpose of the IRB is to review all proposals for human research *before* the research is conducted to ascertain whether the research plan has adequately included the ethical dimensions of the project. The administration of IRBs by the DHHS is conducted by the Office for Protection from Research Risks (OPRR), except for drug-related research, which is administered by the Federal Drug Administration. OPRR is an office within the National Institutes of Health. That office answers any queries from local IRBs, provides information to assist IRBs in their functioning, receives and investigates complaints about research practices, investigates the functioning of local IRBs, as necessary, and recommends sanctions against institutions not in compliance with the law. Institutions not in compliance with the law may lose any federal funding of their programs, including funding of student programs (e.g., federal financial aid to students).

1.3 HOW FEDERAL REGULATIONS AND IRBs CAME ABOUT

Until the past two centuries, people in many cultures considered any kind of research on humans, or even on human cadavers, to be sinful, since they conceived of the spirit and soul as residing in the body.

These religious views about research involving humans largely disappeared with the rise of biomedical research in the 1700s. Research on human subjects gradually gained wide acceptance. However, by the middle of the twentieth century the ethical fallibility of well-meaning scientists was recognized.

The world also came to recognize that atrocities could be committed in the name of science. After World War II, it was learned that Nazi scientists had used prisoners in brutal medical experiments, without the slightest regard for their lives, and had contributed nothing to science in the process. These crimes were investigated at the Nuremberg trials of Nazi war criminals. One consequence of these trials was the development of the Nuremburg Code of research involving humans, which emphasized that scientists must have the informed consent of any human participants in research. Katz (1972) presents a detailed discussion of the origins of political concern about use of human subjects in research.

In the United States, the next significant step in examining research ethics occurred during the 1970s, when the U. S. Congress created the National Commission for the Protection of Human Subjects in Biomedical and Behavioral Research. From 1974 to 1977, the National Commission conducted hearings on ethical problems in human research. On the basis of these hearings and long deliberations, the Commission formulated certain principles and recommendations concerning human research.

The most troubling cases that came to the attention of the National Commission concerned the involvement of human subjects in biomedical research, where concern for human life was sometimes overshadowed by concern for enrolling subjects, completing the research, or using the most rigorous design. To accomplish their scientific objectives, biomedical scientists have at times concealed from subjects circumstances relevant to the subjects' well-being. The following case illustrates how scientific zeal can interfere with ethical sensibilities:

Case 1.2: The Tuskegee Syphilis Study. A study was begun in 1932 to determine the course of syphilis from inception to death. Poor black men were recruited and offered thorough annual examinations and health care in return for serving as subjects in this study. Much information had already been gathered by 1943, when penicillin was identified as a cure for syphilis. However, the subjects were not told of the discovery of an effective treatment for syphilis and the study was allowed to continue until 1972, when an oversight committee finally recognized what was being done and halted the study (Heller, 1972). The details of this case are told in the book *Bad Blood* (Jones, 1982).

The problems that the National Commission observed in the social and behavioral sciences were not of this magnitude, but they were similar in character. In social science research prior to 1973, informed consent was rarely sought and subjects were rarely debriefed or desensitized (restored to an emotional condition at least as good as that with which they had entered the study) after research was performed. In some instances, electric shock was used as a punishing stimulus.

Deception was a standard and unquestioned social research technique, and the assumption seemed to be that subjects neither suspected deception nor could be harmed by it. In retrospect, we see that the harm, while subtle, was manifold. By the 1960s many of the people who participated in research (typically college subjects) actually *expected* deception and produced different results than unsuspecting subjects; see Diener and Crandall (1978, pp. 80-85) for discussion. Naturally, nondeceptive studies also become suspect in the minds of research participants; hence, even the data from studies not employing deception were tainted by the attitudes of subjects expecting to be deceived. Ironically, research validity was being jeopardized by the very procedures thought to promote validity.

Another prevalent problem in the social sciences was invasion of privacy.[5] Social scientists typically study persons who are relatively powerless to refuse (students, the elderly, minority populations), rather than persons who are in a position to limit scientists' access to them—precisely because it is inconvenient, difficult, and even impossible to study the powerful. Like deception, invasion of privacy is not only disrespectful of human subjects but also a cause of invalid data. Those who cannot refuse to participate have a secret weapon available for the protection of their privacy and autonomy—they can lie. Unfortunately, the real harm goes deeper than this apparent game of cat and mouse between investigator and subject. The reputation of social science itself becomes tainted. Consider the following commentary by journalist Nicholas von Hoffman (1970), which appeared in the *Washington Post*:

> We are so preoccupied with defending our privacy against insurance investigators, dope sleuths, counter-espionage men, divorce detectives and credit checkers that we overlook the social scientists behind the hunting blinds who're also peeping into what we thought were our most private and

> secret lives. But there they are, studying us, taking notes, getting to know us, as indifferent as everybody else to the feeling that to be a complete human involves having an aspect of ourselves that's unknown.

Von Hoffman's remarks were about sociologist Laud Humphreys, whose research on "tearoom trade" raises the most difficult of all questions for social scientists: What if there seems to be no way to do an important study without wronging someone?

Case 1.3: Tearoom Trade. The public, as well as law-enforcement authorities, tend to hold simplistic stereotypes about men who commit impersonal sexual acts in public rest rooms. As a consequence, "tearoom sex," as fellatio in public rest rooms is called, used to account for the majority of "homosexual" arrests in the United States. Laud Humphreys, then a doctoral candidate in sociology at Washington University, sought to learn what kinds of men seek quick, impersonal sexual gratification and what motivates them to do so.

Humphreys gathered some of his data by stationing himself in "tearooms" and assuming the role of "watchqueen," the individual who keeps watch and coughs when a police car stops nearby or a stranger approaches. He played that role faithfully while observing hundreds of acts of fellatio. He gained the confidence of some of the men he observed, disclosed to them his role as a scientist, and persuaded them to tell him about the rest of their lives and about their motives for engaging in tearoom trade; but those who were willing to talk openly with him tended to be among the better educated members of the tearoom trade. To avoid socioeconomic class bias, Humphreys secretly followed some of the other men he observed and recorded the license numbers of their cars, which he then surreptitiously matched with Department of Motor Vehicle data to obtain names and addresses. Carefully disguised, Humphreys appeared at their homes a year later and claimed to be a health service interviewer. He interviewed them about their marital status, employment, and so on. Most of these interviews developed into quite personal discussions in which the men disclosed a great deal. Humphreys was aware that his data could be subpoenaed, an eventuality that probably would have led to the arrest of his subjects; he claims to have guarded the data with great care.

Humphreys' findings destroyed stereotypes: Fifty-four percent of his subjects were married and lived with their wives; superficial analysis would suggest that they were exemplary citizens who had satisfactory marriages. Most of these married men did not think of themselves either as bisexual or as homosexual. The marriages of these men were important to them, but were marked with tension. Most of these men or their wives were Catholic, and since the birth of their last child, conjugal relations had been rare, in most cases for reasons connected with family planning. Their alternative source of sexual gratification had to be quick, inexpensive, and impersonal. It could not entail involvement that would threaten their already unstable marriage, or jeopardize their most important asset, their standing as father of their children. They wanted some form of orgasm-producing action that was less lonely than masturbation and less involving than a love relationship. Only about 14% of Humphreys' subjects were members of the gay community and interested primarily in homosexual relations (Humphreys, 1970).

The gay community praised Humphreys' research for dispelling myths and stereotypes. Police departments in some cities responded to the knowledge he produced by ceasing to raid public rest rooms. Many social scientists have applauded Humphreys' research. The Society for the Study of Social Problems chose Humphreys' book for its prestigious annual C. Wright Mills Award. But for others, the study raised some very difficult questions: Is it ever justifiable to act contrary to the interests of subjects in order to obtain valuable knowledge? Does the importance of Humphreys' research justify spying on people and later visiting their homes and families and interviewing them under false pretexts?

Today, a study such as Humphreys' probably would be conducted differently. There are now legal mechanisms for protecting data from subpoena, as well as an emphasis on keeping data in anonymous form if feasible (see Chapter 6). While deception is not entirely ruled out, there is now a strong sentiment against the kinds of deception Humphreys employed (see Chapter 7). For respectful, straightforward approaches to subjects (see Chapter 4). For sensitive use of interview skills to learn about personal matters (see Chapter 11). Using an honest approach, sophisticated interview skills, and assurance of confidentiality, social scientists typically are able to obtain even the most personal information from respondents. Current approaches to research on persons with HIV infection (e.g., Melton, Levine, Koocher, Rosenthal, & Thompson, 1988; see Case 11.1) and on the sexual practices of persons at high risk for AIDS (McKusick, Wiley, & Coates, 1986) attest to the recent advances in social research methodology. Today researchers work with their IRB to develop ethically acceptable procedures.

1.4 HOW IRBs WORK

IRBs consist of five or more members, sometimes including the IRB administrator. The members are required by law to have:

> [V]arying backgrounds to promote complete and adequate review of research activities commonly conducted by the institution. The IRB shall be sufficiently qualified through the experience and expertise of its members, and the diversity of the members' backgrounds including consideration of the racial and cultural backgrounds of members and sensitivity to such issues as community attitudes, to promote respect for its advice and counsel in safeguarding the rights and welfare of human subjects. (45 CFR 46.107, 1981)

The IRB meets periodically to review research protocols submitted by members of that institution and persons intending to do research at that institution. A research protocol of the kind that is submitted to an IRB is a description of the research and of the steps that will be taken to treat subjects respectfully and to reduce any risks involved. See Chapters 2 and 12 for a discussion of protocols. Ideally, the IRB administrator is available to researchers to answer questions and provide information about the IRB. The administrator receives protocols, sends them out for review, calls IRB meetings, and communicates the IRB's concerns or its approvals of protocols.

Human research, as it pertains to IRBs, refers to any study of persons that is a systematic investigation to develop generalizable knowledge. Administrative data gathering that has no scientific purpose is normally not reviewed by IRBs. Classroom demonstrations of research, done solely for pedagogical purposes, are not reviewed by IRBs. There is also a category of exempt research for which the federal government does not insist on IRB review; see Title 45 CFR Part 46.101 for exemptions. Most university IRBs, however, either do not exempt any scientific research from review, or else require that the investigator send a description of the research plan to the IRB to ascertain whether it is, indeed, exempt.

Investigators wishing to do human research should acquaint themselves with the requirements of their IRB at the time they begin planning their research. The IRB administrator can provide investigators with a statement of the requirements of that particular IRB, and of the federal government, concerning human research. While federal regulations outline the general procedures of IRBs, each IRB is responsible for developing its own specific policy statement.

A research protocol may be reviewed by either the full membership or an appropriate subgroup. A protocol must be submitted sufficiently far in advance of scheduled IRB meetings so that members can review it; the IRB administrator can provide information about the review schedule. Those who must begin their research by a particular time should submit their protocols well in advance.

When submitting research proposals under a deadline to a funding agency, it is usually possible to submit a preliminary protocol to the IRB and obtain a letter to the funding agency, stating that the research idea has been approved by the IRB and that a final protocol will be reviewed after the investigator has completed pilot testing and has worked out procedural details. The agency will not release funding until notified by the IRB that the final protocol has been approved.

In the review process, one or more of the reviewers may phone the investigator to clarify questions concerning the protocol. At that time, any problems reviewers have with the protocol can often be resolved. In any event, the investigator will receive a formal letter from the IRB (a) approving the protocol; (b) requesting changes or inquiring about problems; (c) approving the protocol, contingent on the investigator's making specified changes or solving certain problems to the satisfaction of the IRB; or (d) not approving the protocol. Protocols are rarely disapproved outright.

1.5 IS PILOT TESTING REVIEWED BY THE IRB?

Pilot testing refers to informal investigation with one or a few individuals to "fine tune" research procedures until they are satisfactory. For example, when a survey instrument is developed, it typically is tested on a few people and modified various times before it is satisfactory. These people typically are acquaintances of the investigator (e.g., students or colleagues) who have agreed to help with the study. Adequately performed pilot testing also provides an ideal opportunity to discover whether the interests and needs of subjects are adequately met.

Neither fine tuning a questionnaire nor testing equipment or a procedure with the help of a few acquaintances requires IRB review. However, most pilot studies—that is, exploratory studies to determine whether further research might be worthwhile—do require IRB review, as does the pilot testing of a risky procedure. A reviewer in doubt about whether review is required for a pilot activity should check with his or her IRB.

1.6 WHY IRBs HAVE BEEN CONTROVERSIAL

The knowledge required to design research that is both scientifically valid and respectful of human subjects is still not adequately taught in many methodology courses today. Some scientists do not know how to do research that is in compliance with federal regulations. Others find it difficult to describe their research in terms that IRBs readily understand. Not surprisingly, some of these scientists find themselves in an adversarial relationship with their IRB and accuse the federal government of abridging the freedom of science.

IRBs are not perfect, either. An IRB that is unprepared to assist scientists in developing the most acceptable research procedures can only say what is unacceptable—hardly a popular enterprise! When that occurs, the scientist must become an effective ethical problem solver, and be able to communicate about that process with the IRB.

Finally, by establishing a decentralized review system, the federal government has not only given each IRB much autonomy in the interpretation of the regulations but also permits each to add requirements of its own. Thus different IRBs might decide the same case quite differently.

1.7 WHAT IF YOU THINK YOUR IRB MAY DISAPPROVE YOUR PROTOCOL?

Researchers who are aware that their intended research is ethically sensitive must educate themselves about the problems likely to be encountered. They should consult with several sources of information:

1. Scientists who have recently conducted related research.

2. Experts in the pertinent field. For example, if one wants to study the effects of caffeine on various kinds of learning, but realizes that some people have extreme physical reactions to caffeine, the appropriate person to consult may be the campus physician. Similarly, if one wishes to study abused children and is concerned about how they will respond to the intended questionnaire, an appropriate consultant would be a clinical psychologist who treats abused children.

3. Key members of one's own IRB.

4. *IRB, A Review of Human Subjects Research*, an excellent bimonthly journal that covers issues of concern to IRBs and scientists and is available in most university libraries or IRB offices.

In any event, the investigator who undertakes sensitive research must investigate possible risks and learn how to decrease or avoid them. The investigator then describes, in the protocol, the details of the consultation that has occurred, what has been learned about the nature of the possible risks, and what procedures have been selected to minimize those risks. Relevant literature should be discussed and cited.

The IRB may need to be educated. If so, the researcher should provide that needed education and not be adversarial. IRBs have heavy workloads and tire of arrogant colleagues. Besides, they have the last word.

NOTES

1. For the sake of brevity, the term *social research* will be used from here on in place of *social and behavioral research*.

2. Many have argued that the term *research participant* is more respectful than the term *subject*. For some purposes I would agree. For the purposes of this book, however, I would prefer to use a term that continually reminds the reader that the person being studied typically has less power than the researcher and must be accorded the protections that render this inequality morally acceptable.

3. The illustrative cases presented in this book include: (a) cases based on published work; (b) cases based on personal communication; and (c) actual cases known to the author, in which anonymity and deliberate alteration of details are appropriate.

4. The federal regulations governing the protection of human subjects are set forth in Title 45 Code of Federal Regulations (CFR) Part 46. A copy of the federal regulations of human research can be obtained through any university's research office or reference librarian, or from the Office for Protection From Research Risk, National Institutes of Health, Building 31, 9000 Rockville Pike, Bethesda, MD 20892; phone (301) 496-8101.

5. *Privacy* refers to the ability of persons to control intrusions into their personal life; *confidentiality* is an extension of the concept of privacy and refers to agreements governing what may be done with information about oneself. The implications of privacy and confidentiality for research planning are discussed extensively in Chapters 5 and 6.

2

The Research Protocol[1]

This chapter introduces the reader to the concept of the research protocol, and its relation to (a) planning ethically responsible research, (b) working with one's IRB, and (c) using the rest of this book to plan research and develop the protocol. It stresses the importance of using the protocol as a planning tool, not as a bureaucratic evil—a form to be tossed together at the last minute. The details that need to be considered in developing a protocol are discussed in subsequent chapters, and a full discussion of the protocol is presented in Chapter 12.

In research involving humans, as in any complex undertaking, the best way to develop an ethically responsible project is to consider systematically (in writing) the research rationale, methods, and procedures, and the steps that will be taken in response to ethical considerations. Just such a written plan—the protocol—is required by federal regulations of human research. The most effective way to develop an adequate protocol is to begin writing it when the research planning begins; thus, the investigator is reminded to think through the ethical considerations along with the methodological ones. The alternative is to treat ethical considerations as an afterthought and perhaps discover that the research plan is not workable.

2.1 WHAT IS A PROTOCOL?

The research protocol is an official account of the intended research methods and procedures, with special attention to how benefit is maximized and risk minimized, autonomy of subjects is respected, and fairness to subjects is ensured. Included is a brief discussion of the research problem and hypotheses, relevant literature, the research methods, and the investigator's background. This clarifies what is to be done, how, and why. Some other elements of the protocol (and chapters where these are discussed) include:

Subject selection, recruitment, and justification for the number and kind of subjects proposed: Chapters 3, 4, 5, and 12.

Benefits to subjects and others: Chapter 9.

Risks and how these will be minimized, including risks to privacy and confidentiality: Chapters 5, 6, 7, 8, 10, 11, and 12.

Informed consent: Chapters 4, 6, 9, 10, and 12 and Appendix A.

Obtaining permission of a parent or guardian, and subjects' assent, when subjects are minors: Chapter 10.

The protocol might consist of a one- or two-page statement and a consent form, if the project is simple and involves little risk. Or it might be considerably longer. The protocol is prepared by the researcher and submitted to the IRB. It reminds the researcher of many of the elements that are essential to scientifically and ethically sound research, and provides the information needed by an IRB to carry out its legal mandate. A protocol that has been tossed together at the last minute to request IRB approval is likely both to overlook important issues and result in delay of IRB approval.

The protocol enables the investigator and the IRB to ascertain at a glance whether certain matters are handled properly. For example, is the consent statement appropriate? The protocol discusses the purpose and procedures of the research, the characteristics of the research population, the risks and benefits, and the informed consent procedure. Thus, the IRB can observe whether the consent procedure describes the risks and safeguards, the benefits, and the general nature of the research—taking into account the perspective and background of the subjects. A consent statement that overlooks the perspective and background (e.g., culture, education, reading level) of subjects is disrespectful and may adversely affect response rate and cooperation.

The IRB may examine the feasibility of the sampling plan. The protocol states how many subjects will be recruited, from where, and how, and what inducements will be offered to subjects. Does the plan call for too few or too many subjects? Is the subject population suitable to the purpose of the research? Are there concrete plans to benefit those who participate in the research? Is exploitation avoided? In research conducted in an organization (e.g., a school, hospital, workplace, recreation center), the IRB will require the written permission of an authorized gatekeeper for the researcher to approach the subjects. It will require evidence that subjects are not coerced into participating by either the researcher or the gatekeeper. Other things to include depend on the nature of the research.

In a large interview project, one ought to indicate how hired interviewers are trained, and whether they are paid by the hour or "by the head," and why. These matters affect whether subjects are treated respectfully, the success of the sampling procedure, and the validity of the research.

Because the protocol directs the investigator's attention to problems intrinsic to the design and procedure of the research, it is seriously recommended that the investigator begin writing the protocol in the early stages of research planning.

2.2 CONTROL DOCUMENTATION

Institutions are legally responsible for research conducted within them—as are researchers and their supervisors. Therefore, IRB protocols must reflect what is *actually* done in the research. Once the IRB has approved a protocol for a particular project, the investigator is bound to follow that procedure, or to have the desired change of procedure approved by the IRB. *That is, the protocol becomes a control document, an official statement that specifies how the study is being conducted.*

This document becomes a vital part of an official "paper trail" showing that the research is acceptable to a legally constituted board of reviewers. Should anyone raise questions about the project, the approved protocol is powerful evidence that the project is of sufficient value to justify any risks or inconveniences involved.

> *Case 2.1: A (Fictionalized) Study of Moral Development.* Dr. Knowall interviews school children about their understanding of right and wrong. A parent who gave permission for his child to participate in the research later regards the project as seeking to change his child's religious beliefs. He calls the newspaper, the ACLU, the mayor, the school board, and the governor to complain that Dr. Knowall's research violates the separation of church and state. The university is required to respond, and proffers the approved protocol, which would be powerful evidence in any legal proceeding that the project was socially and legally acceptable—

except for one thing: The researcher had slipped in a few questions about religion *after* receiving IRB approval. The researcher finds himself in serious trouble and without enthusiastic backing from his institution.

NOTE

1. The research or treatment protocol is a concept and practice from medicine in which the details of the presenting problem, the patient, and the intended treatment (research) are spelled out in great detail and reviewed by appropriate supervisors to ascertain that it meets the highest ethical, clinical, and research standards. It is developed at the outset, incorporated into the patient's chart, and followed throughout the treatment or research. Unfortunately, when social scientists began to develop protocols for their IRBs, most had no such tradition or training. Rather than use the protocol as a tool for planning and professional consultation, many social scientists regard the protocol as a piece of paperwork one does for the IRB.

3

General Ethical Principles of Research on Humans

The National Commission for the Protection of Human Subjects in Biomedical and Behavioral Research has identified ethical principles and scientific norms that should govern human research. A full discussion appears in *The Belmont Report* (National Commission, 1978). An understanding of these principles and norms will assist the researcher in the planning of research and in the development of the research protocol.

3.1 THREE ETHICAL PRINCIPLES: BENEFICENCE, RESPECT, AND JUSTICE

The following three ethical principles must guide human research:

A. *Beneficence*—maximizing good outcomes for science, humanity, and the individual research participants while avoiding or minimizing unnecessary risk, harm, or wrong.

B. *Respect*—protecting the autonomy of (autonomous) persons, with courtesy and respect for individuals as persons, including those who are not autonomous (e.g., infants, the mentally retarded, senile persons).

C. *Justice*—ensuring reasonable, nonexploitative, and carefully considered procedures and their fair administration; fair distribution of costs and benefits among persons and groups (i.e., those who bear the risks of research should be those who benefit from it).

3.2 SIX NORMS OF SCIENTIFIC RESEARCH

As discussed in *The Belmont Report*, these three basic ethical principles translate into the following six norms of scientific behavior (The letters in parentheses after each norm designate the specific principles upon which that norm is based.):

1. *Valid research design*: Only valid research yields correct results. Valid design takes account of relevant theory, methods, and prior findings; see Chapter 9 for details. (A, B)

2. *Competence of researcher*: The investigator must be capable of carrying out the procedures validly. (A, B)

3. *Identification of consequences*: An assessment of risks and benefits should be identified from relevant perspectives. Ethical research will adjust procedures to respect privacy, ensure confidentiality, maximize benefit, and minimize risk. (A, B, C)

4. *Selection of subjects*: The subjects must be appropriate to the purposes of the study, representative of the population that is to benefit from the research, and appropriate in number; see Chapter 9 for details. (A, B, C)

5. *Voluntary informed consent*: Voluntary informed consent of subjects should be obtained beforehand. *Voluntary* means freely, without threat or undue inducement. *Informed* means that the subject knows what a reasonable person in the same situation would want to know before giving consent. *Consent* means explicit agreement to participate. Informed consent requires clear communication that subjects comprehend, not complex technical explanations or legal jargon. See Chapter 4 for details. (A, B, C)

6. *Compensation for injury*: The researcher is responsible for what happens to subjects. Federal law requires that subjects be informed whether harm will be compensated, but does not require compensation. (A, B, C)

3.3 UNDERSTANDING THE RELATIONSHIP OF NORMS TO PRINCIPLES

These six norms of scientific behavior can be easily derived from the above three ethical principles. The full purpose of these norms cannot be grasped without an understanding of this relationship. To enable the reader to grasp these relationships, an explanation is provided here with respect to the last norm, compensation for injury.

As an exercise, the reader is urged to develop his or her own explanations of the relationships between the three ethical principles and the first five norms and to compare them with the explanations provided in 3.5. Note that these relationships may be explained in various ways.

The norm that the researcher is responsible for compensating participants for injury can best be understood through a simple example: Suppose the researcher intends to interview subjects, who were sexually abused as children, about their perception, as adults, of that experience. This is likely to be traumatic for some. Consequently, the consent statement should indicate (a) the purpose of the research, (b) the possibility that they might experience considerable upset from the interview, and (c) whether they are entitled to psychotherapy from a qualified therapist to work through the immediate sources of their upset.

How does such compensation for injury relate to each ethical principle? The principle of beneficence means maximizing good over harm. We shall assume that the study is well designed and hence may generate useful knowledge. We further assume that participating in the interview may provide an opportunity to recognize, reflect upon, and resolve remaining trauma pertaining to one's prior abuse. But what of the person who cannot resolve the frightening or unpleasant feelings that are generated? Anything that will help that person to reduce the upset engendered by the study would help maximize good over harm. However, psychotherapy for such an individual may take years and require the services of a specialist. Peace of mind might never be restored to some sexually abused individuals. What protection exists against such psychological harm to research subjects? Apart from avoiding trivial research on vulnerable subjects, and offering counseling to those who might need it where affordable to the project, voluntary informed consent provides the main means of protecting subjects.

Respect means both honoring the right of persons to choose whether to be in the study, and showing concern for their well-being. Adults who are competent to consent may choose to be in a study that is potentially upsetting, but the investigator must also respect their well-being. To cause upset and not take appropriate steps to restore the person's sense of well-being is disrespectful. It is unjust that some should be left to suffer as a result of their yielding valuable knowledge that may benefit others. Any steps that can be taken to observe when participants have been upset, and to offer them the resources required to achieve well-being, contribute to the achievement of justice.

3.4 APPLYING THESE PRINCIPLES AND NORMS TO THE DESIGN OF RESEARCH

These norms and principles are powerful tools for identifying ethical issues in research. Practice in applying these principles to some hypothetical research plans will help prepare the reader for the application of these concepts to actual cases. Determine which norms apply to each problem. Evaluate what should be done about each research proposal.

1. A researcher plans to study the effects of competition on ability to solve math problems. Half of the subjects will be told that the researcher wants to see what approach they take in solving math problems. The other half will be told that the researcher wants to see which persons choose the *best* approach.

2. A researcher plans to compare the intellectual skills of retired people to those of college sophomores. To recruit the sophomores, she plans to arrange for volunteers to receive an A in their psychology course, and for nonvolunteers to have their grade lowered. To recruit retired people, she plans to go to a retirement community each evening, knock at people's doors, and ask them to work some puzzles, not explaining details of the study because most wouldn't understand. Because retired people are usually unwilling to participate, only three of them will be recruited.

3. A graduate student plans to compare cocaine use in college freshmen and seniors. Because she may want to reinterview some subjects later, she plans to write their names and phone numbers on their data sheets. She plans to promise confidentiality, so that subjects will trust her, and to keep the data in her dorm room in a locked file.

4. A researcher plans to study the effects of an educational (cable) TV curriculum on learning to read. He gives access to the cable TV programs to 100 homes with 5-year-olds, where the parents agree to allow their children to watch the TV curriculum daily. He obtains permission to test these 100 children in 2 months, along with 100 matched control children who will not have access to the cable TV.

5. To study self-esteem, a researcher plans to have 8-year-olds draw pictures of themselves and their friends and answer some questions. She plans to ask a teacher friend to let her test her students.

3.5 SOME ANSWERS TO THE EXERCISES IN 3.3

Norm 1. Invalid research cannot provide scientifically sound knowledge; in fact, it may provide misleading and thus socially harmful information. Such research does more harm than good and cannot be considered beneficent. It is disrespectful to use people for invalid research.

Norm 2. A researcher who is incompetent to carry out the research cannot produce good outcomes or involve subjects wisely. It is disrespectful to involve subjects in this way.

Norm 3. Beneficence requires discovery of ways to maximize benefits over risks. Research that is respectful of participants does not place them at unnecessary risk. It is unjust to subject persons to unnecessary risk to benefit others.

Norm 4. The validity, hence beneficence, of research depends on selection of the appropriate sample and number of subjects. It is disrespectful and wasteful of participants' time (a) to subject an inappropriate group or inappropriate-size sample to the inconvenience of being studied (for naught), and (b) to subject vulnerable people to procedures to which they are in no position to object. It is unfair to do research on a population that will not benefit from the knowledge gained, in order to benefit some other population. This is justified only when there is no other feasible way to do the research and the benefit to others promises to be great.

Norm 5. Voluntary informed consent is respectful of individuals' autonomy. It is also beneficent in that it gives persons the opportunity to decide for themselves whether participation would be a beneficial experience for them and whether it would involve any risks they are unwilling to take.

3.6 OTHER ETHICAL PRINCIPLES

The principles and norms set forth by the National Commission in *The Belmont Report* were intended to provide succinct guidelines to govern all of biomedical and behavioral research. Other groups (e.g., the American Psychological Association, the American Sociological Association, the American Anthropological Association) have set forth guidelines that are designed to guide research in their respective disciplines. While congruent with the principles and norms set forth by the National Commission, these guidelines explicitly address some additional issues that the National Commission did not mention. Given that each discipline tends to encounter somewhat different ethical problems in research, it is useful for investigators to be familiar with the guidelines intended for their own discipline. (The full text of the professional code of ethics of each association is available from the association office, in Washington, D. C.)

For example, the American Psychological Association (APA) recognizes that the researcher cannot always reveal the exact purpose of the research ahead of time and stresses the importance of debriefing and of removing any undesirable consequences for the individual participant. Information obtained from the research participant is considered confidential unless otherwise agreed upon in advance (APA, 1982).

The American Sociological Association (ASA) recognizes that its researchers often work to influence social policy. Accordingly, emphasis is placed on not misrepresenting one's abilities to conduct a particular research project, honest reporting, disclosure of all sources of financial support and special relations to the sponsor, and nonacceptance of grants that would violate the ASA code of ethics. Sociologists are to lend their expertise on a *pro bono* basis to organizations that cannot afford to fund them. In joint research projects, there should be explicit agreements on division of work, compensation, rights of authorship, access to data, and other rights and responsibilities.

The American Anthropological Association focuses on the context of fieldwork. Anthropologists work throughout the world in close personal association with people in developing countries. Consequently, their relationship with their sponsors, their own and host governments, their students, and the particular individuals and groups they study are unusually complex and fraught with potential for misunderstanding, conflict and possible harm to individuals, groups, and cultures. Anthropologists are exhorted not to pursue a particular piece of research if they cannot do so without damaging those they would study.

Various guidelines and casebooks that have been written for social science disciplines are described below.

RECOMMENDED READINGS

American Psychological Association. (1982). *Ethical principles in the conduct of research with human participants*. Washington, DC: Author. [This book examines the implications of each principle.]

Cassell, J., & Jacobs, S-E. (1987). *Handbook on ethical issues in anthropology*. Washington, DC: American Anthropological Association.

Fluehr-Lobbau, C. (Ed.). (1991). *Ethics and the profession of anthropology: Dialogue for a new era*. Philadelphia: University of Pennsylvania Press.

Keith-Spiegel, P., & Koocher, G. (1985). *Ethics in psychology: Professional standards and cases*. New York: Random House. [This comprehensive book examines a variety of ethical issues ranging from dual role relationships and psychological testing to research issues and mass media audiences.]

Sieber, J. (Ed.). (1982). *Vol. I: The ethics of social research: Surveys and experiments*. New York: Springer-Verlag.

Sieber, J. (Ed.). (1982). *Vol. II: The ethics of social research: Fieldwork, regulation and publication*. New York: Springer-Verlag.

PART II

Basic Ethical Issues in Social and Behavioral Research

The methods and ethics of research may be conceptually distinct topics, but in practice they are inseparable. Poor quality data are obtained when the investigator is insensitive to the needs and interests of subjects. This section answers the following questions:

What is the impact of informed consent on the ability to obtain high quality data? What are the psychological and legal elements of informed consent? How does one obtain consent and debrief subjects in the various settings where research may be performed?

How does an investigator become sensitive to what is private to subjects, and respect those privacies? What is the relationship of respect for privacy to validity of data?

What promises of confidentiality should and should not be made? How can an investigator assure the confidentiality that is promised? How do consent and confidentiality influence the kinds of information subjects are willing to provide to investigators?

When is the use of deception acceptable?

The more general but equally basic ethical question of how to minimize risk and maximize benefit is answered in Part III.

4

Voluntary Informed Consent and Debriefing

Voluntary informed consent is an ongoing, two-way communication process between subjects and the investigator, as well as a specific agreement about the conditions of the research participation. *Voluntary* means without threat or undue inducement. *Informed* means that the subject knows what a reasonable person in the same situation would want to know before giving consent. *Consent* means explicit agreement to participate. Informed consent requires clear communication, not complex technical explanations or legal jargon beyond the subject's ability to comprehend. Social scientists should draw upon their communication skills to ensure that the consent process fulfills these criteria and that communication lines remain open, even after the formal and legally mandated consent has occurred.

We will concentrate first on the communication process of obtaining informed consent, then on the legal requirements, and finally on debriefing. A few issues of consent and debriefing in community-based settings are introduced here, although extensive discussion of these issues is deferred until Chapter 11. Issues of consent for research on children are discussed in Chapter 10, and issues of debriefing in research involving deception are discussed in Chapter 7.

4.1 THE COMMUNICATION PROCESS OF VOLUNTARY INFORMED CONSENT

There are many aspects of the investigator's speech and behavior that communicate information to subjects. Body language, friendliness, a respectful attitude, and genuine empathy for the role of the subject are among the factors that may speak louder than words. To illustrate, imagine a potential subject who is waiting to participate in a study:

Scenario 1: The scientist arrives late, wearing a rumpled lab coat, and props himself in the doorway. He ascertains that the subject is indeed the person whose name is on his list. He reads the consent information without looking at the subject. The subject tries to discuss the information with the researcher, who seems not to hear. He reads off the possible risks. The nonverbal communication that has occurred is powerful. The subject feels resentful and suppresses an urge to storm out. What has been communicated most clearly is that the investigator does not care about the subject. The subject is sophisticated and recognizes that the researcher is immature, preoccupied, and lacking in social skills, yet he feels devalued. He silently succumbs to the pressures of this unequal status relationship to do "the right thing"; he signs the consent form amidst a rush of unpleasant emotions.

Scenario 2: The subject enters the anteroom and meets a researcher who is well groomed, stands straight and relaxed, and invites the subject to sit down with him. The researcher's eye contact, easy and relaxed approach, warm but professional manner, voice, breathing, and a host of other cues convey that he is comfortable communicating with the subject. He is friendly and direct as he describes the study. Through eye contact, he ascertains that the subject understands what he has said. He invites questions, and responds thoughtfully to any comments, questions or concerns. If the subject raises a scientific question about the study (no matter how naive), the scientist welcomes the subject's interest in the project and enters into a brief discussion, treating the subject as a respected peer. Finally, he indicates that there is a formal consent form to be signed and shows the subject that the consent form covers the issues that were discussed. He mentions that it is important that people not feel pressured to participate, but rather participate only if they really want to. The subject signs the form and receives a copy of the form to keep for himself.

Though the consent forms in the first and second case may have been identical, only the second case exemplified adequate, respectful informed consent. In that case, the researcher engendered a strong sense of rapport, trust, and mutual respect; he was responsive to the concerns of the subject and he facilitated adequate decision making. Let us analyze these and other elements of communication:

Rapport. Because informed consent procedures are administered to many subjects in some experiments, it is all too easy to turn the process into a singsong routine that is delivered without any sense of commitment to interpersonal communication. A friendly greeting, openness, positive body language, and a genuine willingness to hear what each subject has to say or ask about the study are crucial to establishing rapport. The amount of eye contact one should employ depends on various circumstances. Extensive eye contact can interfere with the subject's ability to think, and would be considered rude in some Asian cultures. Too little eye contact may signal avoidance, however. Lack of rapport communicates disrespect.

Congruence of verbal and body language. This is an important part of rapport. In the above two examples, the first researcher was highly incongruent: The words said one thing, the manner in which they were delivered said the opposite. The second researcher was highly congruent: All channels of his communication conveyed respect and openness. The congruent communicator of informed consent uses vocabulary that the subject can easily understand, speaks in gentle, direct tones at about the same rate of speech that the subject uses, breathes deeply and calmly, stands or sits straight and relaxed, and is accessible to eye contact. Even if the researcher was feeling stressed, he or she takes time to relax so as not to make distracting movements, show impatience, or laugh inappropriately. To communicate congruently, one's mind must be relatively clear of distracting thoughts.

Trust. If participants believe that the investigator may not understand or care about them, there will not be the sense of partnership needed to carry out the study satisfactorily. The issue of trust is particularly important when the investigator has considerably higher status than members of the target population, or is from a different ethnic group. It is often useful to ask representatives of the subject population to examine the research procedures and make sure they are respectful and acceptable to the target population, as the following example illustrates:

> A Caucasian anthropologist wanted to interview families in San Francisco's Chinatown to determine what kinds of foods they eat,

how their eating habits have changed since they immigrated here, and what incidence of cancer has been experienced in their family. She employed several Chinese-American women to learn whether her interview questions were appropriate and to translate them into Mandarin and Cantonese. First, the research assistants worked on the basis of their personal knowledge of the language and culture of Chinatown; they then tested their procedures on pilot subjects. There was considerable confusion among pilot subjects about the names of some Chinese vegetables; the researchers devised pictures of those vegetables so that subjects could confirm which ones they meant. The Chinese-American research assistants rewrote the questions and the consent statement until they were appropriate for the population that was to be interviewed, then conducted the interviews. Their appearance, language, and cultural background engendered a level of trust, mutual respect, and clear communication that the researcher herself could not have created.

Another way to have built trust and cooperation in that community would have been to identify legitimate leaders or gatekeepers, who are concerned about the health and welfare of community members, and to work with them to make the survey mutually useful. A gatekeeper is a person who lets researchers into the setting or keeps them out. Gatekeepers derive their power from their ability to negotiate conditions that are acceptable to those they serve. Only unscrupulous gatekeepers would grant a researcher privileges that would cause concern or harm to research participants or to the community. Gatekeepers may be scientists, such as a researcher who also directs a clinic. More frequently, they are nonscientists—principals or school-district superintendents, managers of companies, directors of agencies, ministers of local churches, or "street professionals," such as a recovered drug addict who now serves as a community outreach person to his own people.

Some anthropologists have offered to share data with their host community for its own policy-making purposes (e.g., Pelto, 1988; White, 1991). The community leaders or gatekeepers might request that certain items of interest to them be added to a survey and might subsequently need some assistance with specific analyses and interpretations of data. The net result could be a collaborative effort to achieve a shared goal, such as improve health and nutrition in that community. Ideally the collaboration and cooperation would be communicated explicitly to community members. For example, the community newspaper might print an article—including pictures of the interviewers who would soon appear at residents' doors. Interviewers might even carry copies of the newspaper article with them for purposes of identification.

There are many ways to enhance rapport, respect, and trust and increase the benefit to subjects of the research project, depending on the particular setting and circumstances. When planning research, especially in a field setting, it is useful to conduct focus groups from the target population (see Part III, and Stewart & Shamdasani, 1990), to consult with community gatekeepers (Chapter 11), or simply to consult with pilot subjects. The purpose of such consultation during planning is to learn how subjects are likely to react to the various possible research procedures and how to make the research most beneficial and acceptable to subjects. The rewards to the researcher for this effort include greater ease of recruiting, cooperative research participants, a research design that will work, and a community that evinces good will.

Relevance to the concerns of the research population. In developing consent statements, researchers usually try to address the concerns they think their subjects ought to have. However, it is important for the researcher to determine what the concerns of that subject population actually are. *Pilot subjects* from the research population should have the procedure explained to them and should be asked to try to imagine what concerns people would have about participating in the study. Often some of these concerns turn out to be very different from those the researcher would have imagined, and they are likely to affect the outcome of the research if they are not resolved, as the following case illustrates:

> *Case 4.1: Misinformed Consent.* A Ph.D. student interviewed aged persons living in a publicly supported geriatric center on their perceptions of the center. At the time of the research, city budget cuts were occurring; rumors were rampant that eligibility criteria would change and many current residents would be evicted. Mrs. B., an amputee, was fearful that she would be moved if she were perceived as incompetent. Upon signing the informed consent form she began answering his questions:
>
> "Can you recite the alphabet?"
>
> "Backwards or forwards?" she asked to demonstrate her intellectual competence.
>
> "How do you like the service here?"

> "Oh, it's great!" she replied, although she constantly complained to her family about the poor service and bad food.
>
> "How do you like the food here?"
>
> "It's delicious," she replied.
>
> Mrs. B.'s anxiety was rising and midway through the questioning she asked the student, "Did I pass the test?"
>
> "What test?" he asked.
>
> "The one for whether I can stay in the hospital."
>
> "I'm not working for the hospital," he replied. With that, Mrs. B. spun her chair around and wheeled herself away. (Fisher & Rosendahl, 1990, pp. 47-48)

Comprehension. In addition to its relevance to the concerns of the research population, the consent must employ terms and concepts that they will understand. To check for understandability, pilot subjects should be asked to read the consent statement and explain it in their own words. It should be revised until it is correctly understood.

Adequacy of decision making. Even when rapport, comprehension, relevance, and trust are present, it is possible that a subject may fail to give adequate consideration to the decision to participate. Adequate decision making is important to both the subject and the researcher. The subject who regrets agreeing to participate in a study is likely to be late or fail to appear at all, to hurry through the procedures with less than full attention, or even to give dishonest answers.

When consent statements are presented as a plea for help, two factors may cause subjects to participate, even though they would rather not. The *volunteer effect* (Rosenthal & Rosnow, 1969) occurs when subjects feel that they ought to be helpful and agree to participate to do "the right thing." The other factor that predisposes people to be poor decision makers is being rushed into a decision. The following steps will help to avoid these two problems:

1. Present the consent statement well before subjects are to participate, so that they have ample time to consider their decision.
2. Especially if participation requires much time and effort, urge subjects to make the decision that best serves their own interests, as to do otherwise will serve no one's interests.
3. Provide a group context in which subjects discuss with the researcher the pros and cons of participating. This gives individuals exposure to much more information, both for and against participation, than individual decision makers would typically generate.

4. If the procedure is complicated and unusual, let subjects participate in a simulation, or show a videotape of another subject participating, to provide a concrete sense of what is involved.

5. If some or all of the intended subjects do not speak English, the consent statement should be translated by a bilingual person who fully understands the research and the research population. A second bilingual person should then translate the statement back to English to detect any possible misunderstandings in the original translation. Employ the first four procedures above, as appropriate, in the native language of the research population.

Competency and voluntariness in special populations. Although the competence to understand and make decisions about research participation is conceptually distinct from voluntariness, these qualities become blurred in the case of some "captive" research populations. Children, retarded adults, the poorly educated, and prisoners may fail to understand their right to refuse to participate in research when asked by someone of apparent authority. They may also fail to grasp the details relevant to their decision. Where competency is a legal issue, the matter is resolved by appointing an advocate for the research subject, in addition to obtaining the subject's assent. Children cannot legally consent to participate in research, but they can indicate whether they want to participate, and must be given veto power over adults who give permission for them to participate. This is called *assent*. (See Chapter 10 for a detailed discussion of assent and research on children.)

Competence to consent or assent and voluntariness are affected by the way the decision is presented (Melton & Stanley, 1991). For example, an individual's understanding of information presented in the consent procedure, and acceptance of his or her status as an autonomous decision maker, will be most powerfully influenced not by *what* he or she is told but by *how* he or she is engaged in the communication. See Stanley and Guido (1991) for a review of literature on competency and voluntariness.

Protection of privacy and confidentiality.[1] It is essential that researchers protect the privacy of research participants and the confidentiality of data to the extent possible, and communicate how this will be done (including limits on their ability to assure confidentiality) in the consent statement. This is discussed extensively in Chapter 5 (Privacy) and Chapter 6 (Confidentiality).

4.2 LEGAL ELEMENTS OF VOLUNTARY INFORMED CONSENT

Federal law requires that the formal consent statement contain the following information:

1. An explanation of the purpose of the research, the expected duration of the subject's participation, and a description of the procedure. There is no need to describe the details of the design, especially if this will affect the subjects' responses in ways that jeopardize the validity of the research. If concealment is necessary, the subject should be told that not all of the details of the research can be revealed until later, at which time a full explanation will be given. It is necessary only to describe, in terms the subject understands, what the subject will experience. Jargon, legalistic terminology, and irrelevant explanations should be avoided.

2. A description of any foreseeable risk or discomfort. (See Chapter 8 for a full discussion of the meanings of *risk* and how it may be foreseen or assessed.)

3. A description of any benefits to subjects or others reasonably to be expected. (See Chapter 9.)

4. A description of alternatives to participation that might be advantageous to the subject. As examples, subject pool participants must be given viable alternatives to participation (e.g., Sieber & Saks, 1989); persons who have sought clinical treatment and are offered an experimental treatment must be offered the standard treatment as an alternative.

5. A description of how confidentiality or anonymity will be assured and the limits to such assurances if warranted—especially in sensitive research, such as AIDS research. (See 4.4 and Chapter 6 for details.)

6. For research involving more than minimal risk, a statement of whether compensation or treatment for harm or injury is available. This is more relevant to biomedical research in which physical harm is a distinct possibility. In social and behavioral research, the possible harms are more likely to be social and emotional. Thus, for example, if research may be upsetting, the investigator may be available for immediate counseling or may provide the services of an appropriate professional.

7. An explanation of whom to contact for answers to pertinent questions about the research and about subjects' rights, and whom to contact in the event of research-related harm.

8. Indication that participation is voluntary, that refusal to participate will involve no penalty or loss of benefits to which the subject would otherwise be entitled, and that the subject may discontinue participation at any time.

9. The subject should be given a copy of the consent statement.

Other elements that may be appropriate only to medical research are specified in 45 CFR 26.116 of the federal regulations. Common sense dictates that other elements be included in social research. For example, if the research is sponsored (and especially if it is sponsored by an organization that not everyone approves of), the investigator should disclose the identity of the sponsor.

4.3 EFFECTIVE CONSENT STATEMENTS

Although consent statements should explain the research to be undertaken and should fulfill legal requirements, they should also be simple and friendly in tone. Not surprisingly, some consent statements are written in scientific jargon, telling the subject more about the design than the subject cares to know, yet failing to mention things that a person would need to know in order to decide whether to participate. Others are written in harsh, legal-sounding language. An effective consent statement should translate a scientific proposal into simple, everyday language, omitting details that are unimportant to the subjects.

The following exercise provides an opportunity to translate a technical description of a research design into an adequate consent statement. Assume that you are a researcher who has been given permission to study taste perception at a hospital and has designed the following study. You now need to write a clear, friendly letter soliciting participation in the research. Here is the project, as described in scientific language. Translate this into a consent letter that uses plain English:

> Eating disorders (e.g., cravings and aversions) have been observed among psychiatric patients receiving lithium treatment. Acuity for detecting and recognizing the four basic tastes (sweet, sour, salty, and bitter) and preference among these tastes will be measured among patients undergoing lithium therapy and a group of matched controls. It is hypothesized that lithium-medicated subjects have altered taste perception thresholds and taste preferences. The substances to be

tasted will consist of pure water and small concentrations of the following substances diluted in water: sucrose (sweet), salt, citric acid (sour), and quinine sulfate (bitter). These substances are normally used as food additives at higher levels of concentration. Three small samples will be presented simultaneously, two identical and one different, with position varied so that odd and identical samples are tried equally often. Paired comparisons and an hedonic rating scale will be used to measure taste preference. Data acquired through taste testing will be analyzed in relation to age, sex, smoking history, duration of lithium administration, and current lithium concentration. Five sessions per subject, each 10 to 15 minutes long, are required. Three threshold tests for each of the four test substances will be conducted on separate days. On the fifth day, preference testing will be conducted.[2]

Make sure your letter contains all of the required elements of consent, is easy to understand, is friendly, and contains no unnecessary detail or jargon. Remember, you do not need to describe the study as though you were describing it to a scientist; rather, describe generally why the study is being done and what the subject will experience. Be sure to cover all of the elements of consent listed in 4.2. An appropriate letter probably should devote no more than two sentences to each of the following points of information:

1. Identification of the researcher.
2. Explanation of the purpose of the study.
3. Request for participation, mentioning right to withdraw at any time with impunity.
4. Explanation of research method.
5. Duration of research participation.
6. A description of how confidentiality will be maintained.
7. Mention of the subject's right of refusal without penalty.
8. Mention of right to withdraw own data at end of session.
9. Explanation of any risks.
10. Description of any feedback and benefits to subjects.
11. Information on how to contact the person designated to answer questions about subjects' rights or injuries.
12. Indication that subjects may keep a copy of the consent.

After drafting a consent letter, compare your own with the one that follows. Note the simplicity, friendliness, and lack of jargon:

(Letterhead of the Researcher's Institution)
Dear Patient,

I am a psychologist who specializes in the study of taste perception. I am currently working with the staff of your department to see if we can learn ways to enhance your enjoyment of the food served to you here. We need your help in a new study on how sensitive people are to different tastes and which tastes they prefer. The results of this study may help doctors and dietitians, here and at other hospitals, plan diets to improve health, and may add to the understanding of taste perception.

In this study, we will find out how readily persons detect and identify sweet, sour, salty, and bitter tastes, and which tastes are preferred. This information will be analyzed in relation to some information that I am given by the staff physician from participants' medical records about their age, sex, smoking history, duration of lithium administration, and current lithium concentration. Persons participating in this study can expect to spend about 20 minutes on each of five different days. Participants will be asked to taste plain water and samples of water mixed with small amounts of some safe substances that normally are used to season food; they will be asked to answer some questions about how the samples taste and which ones they prefer. There is no foreseeable risk or discomfort. Participants may withdraw their data at the end of their participation if they decide that they didn't want to participate after all.

Participants' identity and personal information will be kept confidential (locked in a file cabinet to which I alone have access) and will be destroyed as soon as the study is completed. The results will be published in a scientific journal. After the study, all participants will be invited to a presentation on how taste perception works. Then each participant will be given the results of his taste test, and an opportunity to sample foods having both typical and increased amounts of the preferred tastes. We hope you will find this information useful to you in seasoning your food in the cafeteria.

Your participation in this study is strictly voluntary. You may withdraw your participation at any time. Your decision as to whether to participate will have no effect on any benefits you now receive or may need to receive in the future from any agency. For answers to questions pertaining to the research, research participants' rights, or in the event of a research-related injury, you may contact me directly, at 555-1212; Dr. John Smith, Director of Research, at 555-1313; or Dr. Mary Doe, Hospital Director, at 555-1414.
Sincerely yours,
Mary Jones, Research Psychologist

Please indicate your consent by signing a copy of this letter and returning it to me. The other copy is for you to keep.

I have read this letter and consent to participate.
Signature:
Date:

Note that this letter not only fulfills the legal requirements for a consent statement but is also clear, friendly, and respectful of the recipient. Irrespective of whether the researcher is a Nobel Prize-winning scientist or a senior in college, the consent letter should treat the potential subject as an equal. Fanfare about the importance of the project is inappropriate. The letter should make an accurate, but brief, statement about the likely scientific value of the research. Information about the scientific legitimacy of the project is conveyed through the name of the research institution on the letterhead and mention of the researcher's official capacity below the signature (e.g., student, Professor of Sociology, Director).

4.4 CONSENT: SIGNED, ORAL, OR BEHAVIORAL?

Signed (or documented) consent proves that consent was obtained. In the case of some risky research, it may even be desirable to also have a witness sign. Most IRBs require signed consent, except in the following situations (as specified in the federal regulations):

1. When signed consent is intrusive or inconvenient and subjects can behaviorally refuse (e.g., by hanging up on a phone interviewer or by throwing out the survey that was received in the mail). In some cases, a letter or statement providing the information required for informed consent may be sufficient. For example, when phone surveys are conducted using random digit dialing, the researcher does not know the name or address of the subject, thus assuring anonymity; the subject, upon learning about the survey from a brief verbal description, can very easily say no, or simply hang up. The process of getting signed consent may be unduly cumbersome for both parties. Similarly, although a survey mailed to respondents should include a cover letter that contains all of the information required for informed consent, it is generally deemed unnecessary for the subject to actually sign and return a consent form. Returning the survey is the subject's way of consenting. Throwing it away is the subject's way of refusing.

2. When signed consent would jeopardize the well-being of subjects. For example, if the research is on criminal behavior, being recruited and agreeing to participate is evidence of one's illegal activity. When the research focuses on illegal or highly stigmatized aspects of the persons being studied, it is not in the subjects' best interest for the researcher to provide a paper trail that reveals the identity of the subjects. Such information could be subpoenaed by a local law enforcement authority (e.g., in connection with a labor dispute, or prosecution of suspected child abusers or drug dealers), or stolen (e.g., by blackmailers), placing subjects at serious risk. Examples of research in which subject lists might be subpoenaed include studies of drug users or drug dealers, prostitutes, child abusers, pornography producers, white-collar criminals, illegal aliens, and so on. Sometimes, certain information about persons, were it to become known, could also lead to personal, social, or economic discrimination, blackmail, or simply to embarrassment and worry. Having AIDS, being gay, having had an abortion, having been exposed to high levels of radiation—all such characterizations are potentially stigmatizing to individuals. Consequently, in research on populations vulnerable to stigmatizing effects, a signed consent form may well pose more danger than protection to the subjects who participate. (Also, see Chapter 6 [6.6] on legal protections of confidentiality.)

However, just because *signed* consent is not required does not mean that consent is not necessary. Consent is necessary and a copy of the consent statement may be given to the subject; only the signed agreement to participate is waived.

4.5 CONSENT AS AN ONGOING PROCESS

The researcher should regard the relationship with subjects as an open communication process and willingly answer questions from subjects and gatekeepers at any time. Especially in field research, where the researcher returns to the site on many occasions, it is easy for his or her welcome to wear thin. In organizational settings, the researcher should provide employees, administrators, and all other interested parties with a written statement (in layperson's language) about the topic of the research and should also furnish occasional updates, reporting on the current phase of work. The research team might also invite members of the organization to drop in at break time for a snack and discussion, questions, or complaints. The advantages of such openness and cordiality are enormous. Even the most sensitive research team gets in the way sometimes and causes unexpected inconvenience to ongoing operations. If people become irritated, but have no avenue for complaining, tensions may grow until they reach a breaking point, and the research is halted. In contrast, openness to any observations and complaints means that problems can be solved before they become serious.

Good communication and good sense require that openness and honesty also characterize the ongoing relationship with subjects, gatekeepers, and others in the research setting. In field settings (e.g., workplaces, schools, hospitals), such communication may occur not only with subjects but also with other key individuals within the setting, such as union representatives, managers, parents, and others who are part of that setting but not subjects. The researcher should be open to discussion of problems that the research raises for the various interests represented within a particular setting. Any communication with the press about the research should be cleared through the gatekeepers of the organization. Sensitivity and willingness to accommodate these interests, however inconvenient for the researcher, pay off in the long run. Insensitivity to such concerns has often resulted in a researcher being asked to leave the field before the project is completed.

4.6 DEBRIEFING AND DISCUSSION OF FINDINGS

"That's all. Thanks. 'Bye." Having gotten the data, the researcher's relationship with subjects has ended. Right? Wrong.

Of paramount importance is *debriefing*, or providing an opportunity for interaction with subjects and relevant others immediately following the research participation. Depending on the nature of the research and the amount of time required to complete the study and analysis, it may also be appropriate to provide an immediate opportunity for *discussion of the findings* of that particular study with subjects and relevant others. Debriefing when deception has occurred is discussed in Chapter 7.

Debriefing. Researchers often state that one benefit of their research is its educational or therapeutic value for participants. The other benefit, of course, is the knowledge gained from subjects. The debriefing process provides an appropriate time to consolidate the educational and therapeutic value to subjects through appropriate conversation and handouts. It is also a good time for the researcher to gain some more knowledge: What were subjects' perceptions of the research? Why did they respond as they did—especially those whose responses were unusual? How do subjects' views of the usefulness of the findings comport with those of the researcher? Typically, the interpretation and application of findings are strengthened by thoughtful discussion with participants. Many a perceptive researcher has learned more from the debriefing process than the data alone could ever reveal.

Debriefing should be a two-way street: Subjects deserve an opportunity to ask questions and express reactions, as well as a few minutes in which to interact with a truly appreciative investigator. The researcher should be listening.

All too often researchers duck their responsibility to debrief by glibly promising to make the results of the study available to subjects. It is rarely a good idea to promise to give subjects the results of the research. The time and logistics involved in doing so can be prohibitive, hence this promise is often broken. Besides, the findings of a single study typically are of limited interest to the participants. Of greater interest and usefulness is the knowledge that the researcher gained from the literature search that (should have) preceded the study, and that can easily be provided during the debriefing. The educational discussion offered after participation should emphasize what is already known, with only secondary attention given to what might be learned from the current study. To do otherwise is to make unwarranted assumptions about the importance of one's research.

The debriefing period should be planned and scheduled as an integral part of the research process. The timing and nature of the debriefing should be appropriate to the circumstances. If the research deals with private or sensitive matters, the debriefing should take place privately with each subject. The complicated study of problem solving in children, described in 4.7, involved four debriefings: (a) several minutes taken with each child after a session, (b) a presentation to the teaching staff at a faculty meeting, (c) a presentation to parents at a PTA meeting, and (d) a brief written presentation mailed to parents. By contrast, in the typical case of research on college students, the debriefing might include a brief, nontechnical presentation on (a) the purpose of the study, (b) the relation of the purpose to the condition(s) in which subjects participated, (c) what is known about the problem and what hypotheses were being tested, (d) the dependent and independent variables, and (e) why the study is of theoretical or practical importance. At this time, subjects also might be given a one- or two-page description of the topic and the research, expanding upon what they had already been told.

The debriefing period is also an appropriate time for such administrative tasks as addressing an envelope for sending a summary of the results when they are ready, paying subjects, and so on.

Discussion of findings. Researchers often appeal for access to a research population on grounds that their research will yield answers that will benefit that population. But after receiving permission to do the research, some researchers forget the implied promise to discuss their findings with those who made the study possible. Publishing the results is not equivalent to first-hand discussion with the research population.

In applied field research in which access is gained through community representatives, arrangements should be made for discussing the findings with appropriate members of the community. Typically, the same gatekeepers who consented to the research decide when, and with whom, the discussion of results should be held. As described in 4.7, preliminary planning of the community presentation occurs as part of the consent process, and final plans are worked out when the presentation of findings is made to the primary gatekeepers. If the results might stigmatize the members of the research population, agreements should be made ahead of time about how such results will be handled; censorship and stigma are to be avoided.

4.7 CONSENT AND DEBRIEFING IN COMMUNITY-BASED RESEARCH

Schools, workplaces, hospitals, prisons, social service agencies, churches, and residential neighborhoods are but a few of the community-based settings where applied research is conducted, and where the researcher must obtain the consent of gatekeepers before seeking the consent of individual subjects. To gain entrée and obtain valid data, the researcher must learn about the culture of the organization in which the research occurs, as well as that of the individuals within it. Thus, the problems discussed in 4.1—the communication elements of consent—

become cross-cultural problems: the scientific and personal culture of the researcher versus the organizational and individual cultures of the gatekeepers and subjects. In some community-based research, such as research on drug addicts at risk for HIV (topics discussed in Chapter 11), the difficulties of comprehending and responding sensitively to these cultures can be overwhelming, even to the experienced community-based researcher. Another problem is that of the unscrupulous or insensitive gatekeeper, who may coerce participation or expect to have access to personal information that the researcher obtains.

The example of consent and debriefing in community-based research presented here is an easy one compared to those discussed in Chapter 11. It is based on my own research on test anxiety and problem solving, conducted in a suburban elementary school on children whose upwardly mobile parents pressured them unduly for good grades. (See Chapter 10 for further discussion of issues of consent for research on children.) Although relatively simple in many respects, the issues of consent and debriefing in this study are similar to most of the questions that must be answered in more difficult settings:

1. *Who is the key gatekeeper? How can needs of the gatekeeper and community be met through the research?* The principal, a mature and understanding woman, had been searching for ways to demonstrate to parents the importance of easing up on their children. She regarded my research as a way to achieve some parent education, provided that I communicated my scientific findings sensitively to parents.

2. *Who are the other gatekeepers? How are they reached?* The school board and superintendent had to approve the research. The teachers had to be satisfied that it would not interfere unduly with their activities. Finally, parents had to consent for their children to participate, and the children, themselves, had to assent. The principal arranged for me to meet with the school board and to attend a teachers' meeting. I drafted a consent letter to parents, and revised it in response to the principal's suggestions.

3. *What flexibility is needed to fit the research plans with the needs of the community?* There was no extra room in the school to accommodate my research; a trailer had to be rented. The schedule, according to which I took students from their classrooms, was worked out in collaboration with their teachers and was subsequently revised as new school events were scheduled. The teachers were concerned that students might think they had not done well on the problem-solving tasks given to them in the

research, so they helped devise the debriefing explanation procedure given to each child after participation.

4. *How should communication and dissemination be handled?* I promised the principal that I would discuss my findings with teachers and parents. The principal subsequently responded to polite inquiries from parents by inviting them to the discussion of the findings at a PTA meeting "sometime in the spring." I reviewed my results and conclusions with the principal, whose astute observations resulted in some constructive changes in my conclusions. I showed her a discussion of my research problem written in layman's language, including a summary of the literature and of my findings, along with some tactful suggestions for helping children overcome test anxiety. After incorporating her suggestions, this paper was sent to parents and teachers, together with a cover letter expressing our gratitude for their cooperation. We also scheduled a presentation of the findings to teachers, and then one to parents at a PTA meeting. I received some nice notes and phone calls from parents and teachers, offering additional comments and thanking me. By the time I wrote the published version of the research, I had gained a perspective that I could not possibly have attained without my conversations with teachers and parents. I also knew that I would be welcome in that school district if I ever wanted to do research there again.

NOTES

1. Throughout this book, *privacy* refers to the interest that persons have in controlling others' access to themselves. *Confidentiality* refers to the agreement between researcher and subject about access by others to the data. *Anonymity* refers to data that include no unique identifiers such as name or Social Security number.

2. I am indebted to Winifred Westberg for this description.

RECOMMENDED READING

Katz, J. (1984). *Silent world of doctor and patient*. New York: Free Press. [Provides excellent illustrations of the process of communication in informed consent.]

5

Privacy

"I know it when I feel it." A gut sense of personal violation may be the tie that binds such disparate events as being subjected to a body search, being the subject of gossip, having one's mail read, being asked one's income, or having one's house entered without permission. It should come as no surprise that such an intensely personal construct is difficult to define.

Melton, 1991, p. 66

We certainly know when our own privacy has been invaded, but do we know when a subject's privacy is likely to be invaded? An interview that asks respondents about matters of love and friendship will produce very different "gut reactions" in someone who has a happy family life versus someone who has just been dumped and is barely controlling feelings of hysteria and despair. How can investigators protect subjects from the pain of invaded privacy? How can investigators guard the integrity of their research against the lies and subterfuges that subjects will employ to hide some private truth or guard against an intrusion?

The purpose of this chapter is to provide a framework for understanding how to respect the privacy of research subjects.

5.1 HOW DO PRIVACY, CONFIDENTIALITY, AND ANONYMITY DIFFER?

Privacy refers to *persons* and to their interest in controlling the access of others to themselves. Confidentiality is an extension of the concept of privacy; it refers to *data* (some record about the person, such as notes or a videotape of the person) and to how data are to be handled in keeping with subjects' interest in controlling the access of others to information about themselves. Ideally, confidentiality is handled in an informed consent agreement between researcher and subject; the agreement states what may be done with private information that the subject conveys to the researcher. The terms of the confidentiality agreement need to be tailored to the particular situation. Anonymity means that the names and other unique identifiers (e.g., Social Security number, address) of subjects are never attached to the data or even known to the researcher.

5.2 WHY IS PRIVACY AN ISSUE IN RESEARCH?

Privacy, that is, subjects' degree of control of the access that others have to them and to information about them, affects their willingness to participate in research and to give honest responses. An understanding of the privacy concerns of potential subjects enables the researcher to communicate an awareness of and respect for those concerns, and to protect subjects from invasion of their privacy. Because privacy issues are often subtle and not understood by the researcher, appropriate awareness and safeguards may be omitted, with unfortunate results. Here are some examples:

> A researcher interviews poor Chicano families in Los Angeles about their attitudes concerning AIDS. Unknown to the researcher, these people consider it immoral, and sacrilegious, to even talk about homosexuality or AIDS. Most pretend not to understand his questions.
>
> A researcher gets access to medical records, discovers which persons have asthma, and contacts them directly to ask them to participate in research on coping strategies of asthmatics. "How did you get my name?" "What were you doing with my medical records?" were the thoughts, if not the actual questions, of most of those called. Most refused to participate. The researcher should have asked physicians to send their asthmatic patients a letter (drafted and paid for by the researcher), asking if they would be interested in participating, and if so, the physician would release their names to the researcher.
>
> A researcher interviews families about their child-rearing practices. She establishes such excellent rapport that some pour out details of

physical and sexual abuse of children by other members of their families, all of which the researcher is required by law to report, even though she has promised confidentiality.

A researcher interviews children about their moral beliefs. Believing that the children would want privacy, he interviews 5-year-olds alone. However, they are sufficiently afraid to be alone that they do not respond as well as they would have if their mothers had been present. Recognizing his error, the researcher then makes sure that subjects from the next group, 8-year-olds, are accompanied by their mothers. However, the 8-year-olds have entered that stage of development in which some privacy from parents is important. Consequently, they do not answer all of his questions honestly.

A researcher decides to use telephone interviews to learn about the health history of lower-class older people, as the phone typically offers greater privacy than the face-to-face interview. She fails to recognize, however, that poor elderly people rarely live alone or have privacy from their families when they use the phone, and many keep health secrets from their family.

A researcher interviews minority children about their nutritional habits. Because another scientist has made known to the press the opinion that this community does not responsibly feed its children, a small group of activists in the community breaks into the researcher's files and steals his data.

In each case, the researcher has been insensitive to privacy issues idiosyncratic to the research population and has not addressed the problems these issues pose for his or her research. Had he or she consulted with community gatekeepers or others familiar with the research population, these problems might have been identified and solved. Most of the research topics that interest social scientists concern somewhat private or personal matters. Yet most topics, however private, can be effectively and responsibly researched if the researcher employs appropriate sensitivity and safeguards.

5.3 IS THERE A RIGHT TO PRIVACY?

An individual's right to privacy from research inquiry is protected by the right to refuse to participate in research. An investigator is free to do research on consenting subjects or on publicly available information, including unobtrusive observation of people in public places. May a researcher videotape or photograph behavior in public without obtaining informed consent? There is no law against this, but common courtesy and sensitivity to local norms should be heeded. Intimate acts, such as goodbyes at airports, should be regarded as private, even though performed in public.

Constitutional and federal laws have little to say directly about privacy and social research. The only definitive federal privacy laws governing social research are the following: The Buckley Amendment prohibits access to children's school records without parental consent. The Hatch Act prohibits asking children questions about religion, sex, or family life without parental permission. And the National Research Act requires parental permission for research on children.

Tort law provides a mechanism under which persons might take action against an investigator alleged to have invaded privacy. In such an action, the law defines privacy in relation to other interests. It expects behavioral scientists to be sensitive to persons' claims to privacy, but recognizes that claims to privacy must sometimes yield to competing claims. Any subject may file a suit against a researcher for "invasion of privacy," but courts of law are sensitive to the value of research as well as the value of privacy.

An important protection against such a suit is an adequate informed consent statement signed by each participant, as well as parental permission for research participation by children. However, persons other than the research participant may consider their privacy invaded by the research. For example, family members who are not participants in the research may feel that the investigation also probed their affairs. If the research is socially important and validly designed, if the researcher has taken reasonable precautions to respect the privacy needs of typical subjects and others associated with the research, and if the project has been approved by an IRB, such a suit is likely to be dismissed. But what exactly is this privacy about which researchers need to be so careful?

5.4 A BEHAVIORAL DEFINITION OF PRIVACY

As a behavioral phenomenon, privacy refers to certain needs to establish personal boundaries; these needs seem to be basic and universal, but they are manifested differently, depending on learning and cultural and developmental factors. Privacy does not simply mean being left alone. Some people have too little opportunity to share life with others, or to bask in public attention. When treated respectfully, many are pleased that an investigator is interested in hearing about their personal lives. Because of this desire on the part of lonely people for understanding and attention, competent survey investigators often have more difficulty exiting people's homes than entering.

A different kind of unwanted privacy was found by Klockars (1974), a criminologist, when he undertook a case study of a well-known "fence." The fence was an elderly pawnshop owner, who had stolen or fenced vast amounts of goods in the course of his life. Klockars told the fence that he would like to document the details of his career, as the world has little biographical information about the lives of famous thieves. The fence wanted to go down in history and offered to tell all, provided that Klockars used the fence's real name and address in his writing and published the entire account in a book (Klockars, 1974). This was done, and the aging fence proudly decorated his pawnshop with clippings from the book.

Privacy is invaded when people are given unwanted information: A subject's privacy may be breached by showing him pornography, or requiring him to listen to more about some other person's sex life than he would care to hear. Privacy is also invaded when people are deprived of their normal flow of information, as when unconsenting subjects (who did not realize they were participating in a study) were deprived of information that they would ordinarily use to make important decisions.

Thus, many claims to privacy are also claims to autonomy. For example, subjects' privacy and autonomy are violated when their self-report data on marijuana use become the basis for their arrest, when IQ data are disclosed to school teachers who would use it to track students, or when organizational research data disclosed to managers become the basis for firing or transferring employees. The most dramatic cases in which invasion of privacy results in lowered autonomy are those in which something is done to one's thought processes—the most private part of oneself—through behavior control techniques: for example, psychopharmacology, brainwashing, and subliminal advertising.

5.5 PRIVACY AND INFORMED CONSENT

A research experience regarded by some as a delightful opportunity for self-disclosure could constitute an unbearable invasion of privacy for others. Informed consent is an important way to respect these individual differences. The investigator specifies the kinds of things that will occur in the study, the kinds of information that will be sought and given, and the procedures that will be used to assure anonymity or confidentiality. The subject then decides whether to participate under these conditions. One who considers a given research procedure an invasion of privacy can simply decline to participate.

However, informed consent is not the entire solution. One who is insensitive to the privacy needs of the research population may be unprepared to offer the subjects the forms of respect and protection they want.

5.6 GAINING SENSITIVITY TO PRIVACY INTERESTS OF SUBJECTS

Although there is no way to be sure of the privacy interests of all members of a given research population, the researcher can learn how typical members would feel. If the typical member considers the research activity an invasion of privacy, the data will be badly flawed. Evasion, lying, and dropping out of the study are sure to occur, and those who answer honestly will worry about the consequences.

The best ways to learn about the privacy interests of your research population are as follows: (1) Ask someone who works with that population regularly. For example, ask teachers and parents about the privacy interests of their children, ask a child psychotherapist about the privacy interests of abused children, ask a social worker about the privacy interests of low socioeconomic-status parents; (2) ask an investigator who has had much experience working with that same population; and (3) ask members of the population what they think other people in their group might consider private in relation to the intended study.

5.7 "BROKERED" DATA

Researchers may lack access to a given population for a variety of privacy-related reasons. For example, a researcher may wish to survey people who go to an HIV testing clinic, but the clinic may not want him or her on the premises since his or her presence would reduce clients' sense of privacy and frighten some away. However, the physicians who administer the tests may be willing to hand an anonymous questionnaire to the clients and ask them if they would be willing to respond and mail the questionnaire back at their convenience. The investigator may have to offer some incentive, such as a financial contribution to the clinic, to achieve the necessary cooperation from its staff.

The term *broker* refers to any person who works in some trusted capacity with a population to which the researcher does not have access, and who obtains data from that population for a researcher. For example, the broker may be a psychotherapist or a physician who asks patients if they would provide data for important research being conducted elsewhere. A broker may serve other functions in addition to gathering data for the researcher:

"Broker sanitized" responses. There may be concern that some aspect of the response will enable the investigator to deduce the identity of the respondent. For example, if a survey is sent to organization leaders in various parts of the country, the postmark on the envelope might enable someone to deduce the identity of some respondents. To prevent this, a mutually agreed upon third party may receive all of the responses, remove and destroy the envelopes, then send the responses to the investigator.

Brokers and aliases. Sometimes lists of potential respondents are unavailable directly to the researcher. For example, the researcher wishing to study the attitudes of psychiatric patients at various stages of their therapy may not be privy to their names. Rather, the treating psychiatrists may agree to serve as brokers. The psychiatrists then obtain the informed consent of their patients and periodically gather data from those who consent. Each patient is given an alias. Each time data are gathered, the psychiatrist refers to a list for the alias, substitutes it for the patient's real name, and transmits the completed questionnaire back to the researcher.

Additional roles for the broker. The broker may (a) examine responses for information that might permit deductive disclosure of the identity of the respondent, and remove that information; (b) add information (e.g., a professional evaluation of the respondent); or (c) check responses for accuracy or completeness.

RECOMMENDED READINGS

Laufer, R. S., & Wolfe, M. (1977). Privacy as a concept and a social issue: A multidimensional developmental theory. *Journal of Social Issues, 33*, 44-87. [This elegant theory provides ways of understanding what privacy means to persons in age groups, cultures, and circumstances different from one's own.]

Melton, G. B. (1990). Brief research report: Certificate of confidentiality under the Public Health Service Act: Strong protection but not enough. *Violence and Victims, 5*(1), 67-70. [This paper outlines advantages and problems in the use of the certificate of confidentiality.]

6

Strategies for Assuring Confidentiality

The terms *privacy* and *confidentiality* are often used as though they were interchangeable. They are not. Close attention to the definition of confidentiality is important.

6.1 WHAT IS CONFIDENTIALITY?

Confidentiality refers to agreements with persons about what may be done with their data. The confidentiality agreement between a researcher and subject is part of the informed consent agreement. For example, the following is a confidentiality agreement that might be included in the consent letter of a scientist (or thesis student) seeking to interview families in counseling:

> To protect your privacy, the following measures will ensure that others do not learn your identity or what you tell me.
>
> 1. No names will be used in transcribing from the audio tape, or in writing up the case study. Each person will be assigned a letter name as follows: M for mother, F for father, MS1 for male first sibling, and so on.
> 2. All identifying characteristics, such as occupation, city, and ethnic background, will be changed.
> 3. The audio tapes will be reviewed only in my home (and in the office of my thesis adviser).
> 4. The tapes and notes will be destroyed after my report of this research has been accepted for publication (or in the case of an unpublished thesis—after my thesis has been accepted by the university).

5. What is discussed during our session will be kept confidential with two exceptions: I am compelled by law to inform an appropriate other person if I hear and believe that you are in danger of hurting yourself or someone else, or if there is reasonable suspicion that a child, elder, or dependent adult has been abused.[1]

Noteworthy characteristics of this agreement are: (a) that it recognizes the privacy of some of the information likely to be conveyed; (b) that it states what steps will be taken to ensure that others are not privy to the identity of subjects or to identifiable details about individuals; and (c) that it states legal limitations to the assurance of confidentiality.

6.2 WHY IS CONFIDENTIALITY AN ISSUE IN RESEARCH?

Subjects may be willing to share highly personal information with a researcher if there is a believable confidentiality statement, or if the data are anonymous. Unfortunately, many researchers make glib promises of confidentiality without understanding the ways in which confidentiality may be breached. Subjects, acting on a belief in the researcher's ability to keep the promise, may then be harmed by unintended disclosures. The following examples of broken promises are fictionalized accounts of leaks that have actually occurred in research:

Case 6.1: Blackmail. A researcher studied attitudes concerning morality and asked questions, such as whether subjects had cheated on their income tax, used illegal drugs, had extramarital affairs, or filched supplies from their employers. He also gathered data on the attitudes these persons had expressed at an adult fellowship meeting at their church. He entered each data set into a mainframe computer file, identifying subjects by number. In a separate computer file of that mainframe computer, he kept the linkage of names and numbers. A computer hacker accessed his files. Several subjects were blackmailed. [When storing data on computers to which others have access, identifiers must be stored elsewhere, such as in a safe deposit box.]

Case 6.2: Personnel Action. In a study involving about 200 middle management employees at a large firm, a researcher collected information on drug abuse and financial difficulties, along with employment histories and demographic information such as race, age, and sex. Although no names were used in the final report, it was possible to deduce from the table of summary statistics that a particular employee was a cocaine addict who was about to lose his home because of financial problems. This deduction was possible because he was the only Asian male who had worked continuously in the same division of the company for more than 5 years. He was laid off during the next reduction in work force. [Provide only those summary tables that serve a purpose, and make categories broad enough to prevent singling out of individuals.]

Case 6.3: Subpoena. A researcher interviewed women who had taken medication that was later found to cause birth defects. One of the interview questions was whether respondents remembered reading the warnings of possible side effects in the literature accompanying the drug. In connection with a multimillion-dollar lawsuit against the drug company, the data were subpoenaed to prove that some of the plaintiffs had been aware of the risk they were taking, contrary to what had been stated in the class action suit against the drug company. [This problem would not be solved by destroying unique identifiers. One must be sensitive to when data might be of interest in a law suit. That survey question should never have been asked.]

Case 6.4: Incest. For her master's thesis, a social work student did a case study of a family that had been part of her client load. Fifteen years later, the youngest son in that family entered college and, in connection with a sociology assignment, read the thesis, recognized that it was about his family, and learned that he was actually the son of his eldest sister and their father. [When reporting case studies, the names of persons, places, special events, occupations, ethnic background, and so on should be changed. Any special characteristic of subjects should be changed slightly so that individuals cannot be identified.]

Case 6.5: Computer Problems. A part-time MBA student received permission from the president of the company where she worked to study employee morale while office automation was being introduced. Employees were candid with the researcher because she promised confidentiality and was their trusted co-worker. Some employees complained bitterly to her about the automation system, indicating their conviction that it would fail. She kept the data locked in her office at work, an office to which top executives had skeleton keys. The persons who had complained most bitterly were laid off during the next reduction in staff. [The data should have been kept at home, and codes should have been used, with the code key kept separately.]

In each case, the researcher was responsible for the harm to participants and might have been sued by them. The investigators should have assessed these risks and taken effective precautions. In assessing risk to confidentiality, it is important to remember that attitudes change; the political pendulum swings between conservative and liberal extremes. Consider the following:

Case 6.6: Seizure of Records. (This hypothetical case is based on a police seizure of biomedical research records thought to contain information that would assist in the identification of a bank robber.) A researcher began gathering longitudinal data on the life-style of persons known to be HIV-positive. The data were kept in locked files in a locked office. Three years later, during a conservative political era, a sheriff seeking information on suspected drug dealers issued a search warrant, conducted a "midnight raid," handcuffed the investigator to his desk, and removed his AIDS research files. [Because this is longitudinal research, unique identifiers are needed. When the political climate changed, the researcher should have removed all real names from records in this country and sent the original files to a colleague in another country.]

Adequate safeguards of confidentiality must be employed and described in specific terms in the consent statement. Many people, especially members of minority populations (Turner, 1982), doubt such promises unless the details are spelled out clearly.

6.3 CONFIDENTIALITY OR ANONYMITY?

Anonymity means that the researcher acquires no unique identifiers, such as the subject's name, Social Security number, or driver's license number. When designing the research, one should decide whether the data can be gathered anonymously. Five major reasons for gathering unique identifiers are as follows:

1. So that subjects can be recontacted if their data indicate that they need help or information.

2. So that data sets from the same individual can be linked to one another. (This problem might be solved with code names.)

3. So that results can be mailed to the subject. (This problem can be solved by having each subject address an envelope to himself or herself. Envelopes are stored apart from the data. After results are mailed out, no record of the names of subjects remains.)

4. So that signatures may be obtained on the consent form. (Signed consent may be waived when signatures pose a risk to subjects, as when subjects have been selected because they have engaged in some illegal or socially stigmatizing activity.)

5. So that a low base-rate sample can be identified when a large sample is screened on some measures.

Note that for the first two reasons, the issue is whether to have names associated with subjects' data; for the second two reasons, the issue is whether to have names on file at all. In the fifth case, identifiers may be expunged from the succeeding study as soon as those data are gathered. If the data can be gathered anonymously, subjects will be more forthcoming, and the researcher will be relieved of some responsibilities connected with assuring confidentiality. If the research cannot be done anonymously, the researcher must consider procedural, statistical, and legal methods for assuring confidentiality. Readers who want to investigate methods beyond those presented here are referred to *Assuring the Confidentiality of Social Research Data* (Boruch & Cecil, 1979).

6.4 PROCEDURAL APPROACHES TO ASSURING CONFIDENTIALITY

Certain procedural approaches eliminate or minimize the link between the identifiers and the data. Various procedures are appropriate, depending on whether the research is cross-sectional, longitudinal, or experimental.

6.4.1 Cross-Sectional Research

In cross-sectional research, there is no attempt to link individual data gathered at one time to data gathered at another. Three simple methods of preventing disclosure of unique identifiers in cross-sectional research are as follows:

Anonymity. The researcher has no record of the identity of the respondents. For example, have respondents mail back their questionnaires or hand them back in a group, without names or other unique identifiers.

Temporarily identified responses. It is sometimes important to ensure that only the appropriate persons have responded and that their responses are complete. After checking names against a list or making sure responses are complete, the names are destroyed, as is done at polling places.

Separately identified responses. In mail surveys, it is sometimes necessary to know who has responded and who has not. To accomplish this with an anonymous survey, respondents may be asked to mail back the completed survey anonymously, and to mail separately a postcard with his or her name on it. This method enables the researcher to check off those who have responded and to send another wave of questionnaires to those who have not.

Any of these three methods could be put to corrupt use if the researcher were so inclined. Because people are sensitive to corrupt practices, the honest researcher must demonstrate integrity. The researcher's good name and that of the research institution may reduce suspicion.

6.4.2 Longitudinal Data

In longitudinal studies, one must somehow link together the various responses of a given person. Boruch and Cecil (1979) suggest many approaches, including the following:

Aliases. Subjects use an easily remembered code, such as their mother's birth date, as an alias. The researcher makes sure there are not duplicate codes among respondents. Note that the adequacy of this method depends upon subjects' ability to remember an alias. Inner-city drug addicts are an example of a population that may not remember; in cases where remembering the wrong alias might seriously affect the research or the subject (e.g., the subject gets back the wrong HIV test result), this method of linking data would be clearly inappropriate.

6.4.3 Interfile Linkage

Sometimes a researcher wants to link research records on persons with some other independently stored records on those same persons (exact matching) or else on persons who are similar on some attributes (statistical matching). An example would be court-mandated research on the relationship between academic accomplishment and subsequent arrest records of juveniles who have been sentenced to one of three experimental rehabilitation programs. The court may be unwilling to grant the researcher access to the records involved, but may be willing to arrange for a court clerk to gather all of the relevant data on each subject, then remove identifiers, and give the anonymous files to the researcher. The obvious advantages of exact matching are the ability to obtain data that would be difficult or impossible to obtain otherwise and the ability to construct a longitudinal file.

Statistical matching enables the researcher to create matched comparison groups. An example of statistical matching would be if each boy having a certain test profile were matched with a girl having a similar test profile. Statistical matching may also permit imputation—estimation of the values of missing data—by revealing how similar persons would answer the item. Interfile linkage is a complex set of techniques, the details of which are beyond the purview of this chapter. The interested reader is referred to Campbell, Boruch, Schwartz, and Steinberg (1977) and to Cox and Boruch (1986).

6.5 STATISTICAL STRATEGIES FOR ASSURING CONFIDENTIALITY

It is folly to ask respondents directly if they have engaged in illegal behavior—for example, used cocaine, beaten their children, or cheated on taxes. Respondents are unlikely to answer honestly, and if they did, they would place themselves in legal jeopardy.

Methods of randomized response, or error inoculation, provide a strategy for asking questions in such a way that no one can know who has given incriminating responses. Because they decrease effective sample size and do not work with all populations, these methods should be used only when necessary and appropriate. This procedure is intended to assure subjects, but it arouses suspicion in some populations and diminishes truthfulness.

The simplest variant of this strategy is to give each subject a die to roll before answering a "yes" or "no" question. The respondent might be instructed to answer untruthfully if the die comes up, say, two. Otherwise, he is to answer truthfully. The respondent does not let the researcher see how the die comes up. He then gives his answer according to instructions. The researcher knows that one response in six is false. The following example shows how the data are analyzed:

Suppose the researcher interviews 100 people and asks each if he or she has cheated on income tax; 36 indicate that they have cheated. The following equation enables one to estimate the true proportion of persons who have cheated on their tax:

$$\text{Estimated true proportion} = (Py - Pfp)/(1 - Pfp - Pfn).$$
$$\text{Where } Py = \text{observed proportion of "yes" responses}$$
$$Pfp = \text{probability of false positive responses (here } 1/6)$$
$$Pfn = \text{probability of false negative responses (here } 1/6)$$
$$\text{Estimated true proportion} = (.36 - .167)/(1 - .167 - .167)$$
$$= .193/.666 = .29$$

An estimated 29% of this sample have cheated on their income tax.

There are many variations of this procedure, and a considerable literature on its efficacy. See Boruch and Cecil (1979, 1982) or Fox and Tracy (1986) for details.

6.6 CERTIFICATES OF CONFIDENTIALITY

Members of certain professions, such as priests, physicians, and lawyers, have testimonial privilege. That is, under certain circumstances they may not be required to reveal to a court of law the identity of their clients or sources of information. This privilege does *not* extend to researchers. As indicated earlier, prosecutors, grand juries, legislative bodies, civil litigants, and administrative agencies can use their subpoena powers to compel disclosure of confidential research information. What is to protect the researcher and subjects from this intrusion? Anonymous data, aliases, colleagues in foreign countries, and statistical strategies are not always satisfactory solutions. The most effective and yet underutilized protection against subpoena is the certificate of confidentiality.

In 1988 Congress enacted a law providing for an apparently absolute researcher-participant privilege when it is covered by a certificate of confidentiality issued by the Department of Health and Human Services. The provisions of this relatively new law authorize:

> [P]ersons engaged in biomedical, behavioral, clinical or other research (including research on the use and effect of alcohol and other psychoactive drugs) to protect the privacy of individuals who are the subject of such research by withholding from all persons not connected with the conduct of such research the names, or other identifying characteristics of such individuals. Persons so authorized to protect the privacy of such individuals may not be compelled in any Federal, State, or local civil, criminal administrative, legislative, or other proceedings to identify such individuals. (Public Health Service Act, 301(d), 42 USC 242a)

Certificates of confidentiality are granted on request for any *bona fide* research project of a sensitive nature, in which protection of confidentiality is judged necessary to achieve the research objectives. The research need not be funded or connected with any federal agency. Some government funders also issue certificates of confidentiality to their grantees upon request. Persons interested in learning more about certificates of confidentiality are referred to the Office for Protection from Research Risks, NIH; phone (301) 496-8101; or to the Office of Health Planning and Evaluation, Office of the Assistant Secretary of Health; phone (301) 472-7911.

6.7 CONFIDENTIALITY AND CONSENT

An adequate consent statement shows the subject that the researcher has conducted a thorough analysis of the risks to confidentiality and has acted with the well-being of the subject foremost in mind. The consent statement must specify any promises of confidentiality that the researcher *cannot* make. Typically, these have to do with reporting laws pertaining to child abuse, child molestation, and threats of harm to others. Reporting laws vary from state to state, so the researcher should be familiar with laws in the state where the research is to be conducted. Thus, the consent statement warns the subject not to reveal certain kinds of information to the researcher. A skilled researcher can establish rapport and convince subjects to reveal almost anything, including things the researcher may not want to be responsible for knowing.

There are many ways in which confidentiality might be discussed in a consent statement. A few examples are as follows:

Example 1. To protect your privacy, this research is conducted anonymously. No record of your participation will be kept. Do not sign this consent or put your name on the survey.

Example 2. This is an anonymous study of teacher attitudes and achievements. No names of people, schools, or districts will be gathered. The results will be reported in the form of statistical summaries of group results.

Example 3. The data will be anonymous. You are asked to write your name on the cover sheet because it is essential that I make sure your responses are complete. As soon as you hand in your questionnaire, I will check your responses for completeness and ask you to complete any incomplete items. I will then tear off and destroy the cover sheet. There will be no way anyone else can associate your name with your data.

Example 4. This survey is anonymous. Please complete it and return it unsigned in the enclosed, postage-paid envelope. At the same time, please return the postcard bearing your name. That way we will know you responded, but we will not know which survey is yours.

Example 5. This anonymous study of persons who have decided to be tested for HIV infection is being conducted by Dr. Jan Smith, at Newton University. Because we do not want to intrude on your privacy in any way, Dr. Barry Wray, at the AIDS Testing Center, has agreed to ask you if you would be willing to respond to this survey. Please look it over. If you think you would be willing to respond, take it home, answer the questions, and mail it back to me in the attached, stamped, self-addressed envelope. If you are interested in knowing the results of the study, please write to me at the above address, or stop by the AIDS Testing Center and ask for a copy of the results, which will be available after May 1.

Example 6. Because this is a study in which we hope to track your progress in coping with an incurable disease and your responses to psychotherapy designed to help you in that effort, we will need to interview you every 2 months and match your new interview data with your prior data. To keep your file strictly anonymous, we need to give you an alias. Think of one or more code names you might like to use. Make sure it is a name you will remember, such as the name of a close high school friend, a pet, or a favorite movie star. You will need to check with the researcher to make sure that no other participant has chosen the same name. The name you choose will be the only name that is ever associated with your file. We will be unable to contact you, so we hope you will be sure to keep in touch with us. If you decide to drop out of the study, we would be grateful if you would let us know.

Example 7. In this study, I will examine the relationship between your child's SAT scores and his attitude toward specific areas of study. I respect the privacy of your child. If you give me permission to do so, I will ask your child to fill out an attitude survey. I will then give that survey to the school secretary, who will write your child's SAT subscores on it and erase your child's name from it. That way, I will have attitude and SAT data for each child, but will not know the names of any child. The data will then be statistically analyzed and reported as group data.

These are merely examples. Careful consideration needs to be given to the content and wording of each consent statement.

6.8 DATA SHARING

If research is published, the investigator is accountable for the results and is normally required to keep the data for 5 to 10 years. The editor may ask to see the raw data to check its veracity. Some funders require that the documented data be archived in user-friendly form and made available to other scientists. Data sharing, if done with due respect for confidentiality, is regarded positively by most subjects, who would prefer to think of their data as a contribution to science and available to other legitimate scientists to examine, critique, and build upon.

When data are shared via a public archive, all identifiers must be removed, and the researcher must ensure that there is no way to deduce identity. For example, if data about teachers reveal the name of the district, along with teacher age, sex, and years of teaching, it may be easy for someone with access to school personnel records to deduce identities.

Techniques for rendering data immune to deductive disclosure. A variety of techniques have been developed to transform raw data into a form that prevents deductive disclosure. Variables or cases with easily identifiable characteristics are removed. Random error can be implanted into the data, introducing enough noise to foil attempts at deductive disclosure, but not enough to obscure conclusions. Micro-aggregation creates synthetic individuals; instead of releasing the individual data on 2,000 participants in a study of small business owners, one might group the data into 500 sets of four subjects each, and release average data on every variable for each set, along with the within-variance data. Outside users could do secondary analyses on these 500 synthetic small business owners. For details, see Gates (1988), Kim (1986), Duncan and Lambert (1987), and Boruch and Cecil (1979).

NOTE

1. This was adapted from a statement developed by David H. Ruja, and is discussed in E. Gil. *The California Child Abuse Reporting Law: Issues and Answers for Professionals* [Publication 132(10/86)]. This booklet is printed and distributed by the State of California Department of Social Services, Office of Child Abuse Prevention, 744 P Street, M.S.9-100, Sacramento, CA 95814.

RECOMMENDED READINGS

Boruch R. F., & Cecil, J. S. (1979). *Assuring the confidentiality of social research data.* Philadelphia: University of Pennsylvania Press.

Campbell, D. T., Boruch, R. F., Schwartz, R. D., & Steinberg, J. (1977). Confidentiality-preserving modes of access to files and to interfile exchange for useful statistical analysis. *Evaluation Quarterly, 1*(2), 269-300.

7

Deception Research

In deception research, the researcher studies reactions of subjects who are purposely led to have false beliefs or assumptions. Although deception is essential for the study of some kinds of behavior, there are serious objections to its use. Deception research may deny subjects their right of self-determination, causing a generalized suspicion of research and disrespect for science. It may invade privacy; it may use powerful methods to induce people to do things they regret doing; and it may create a great deal of upset on the part of subjects and society. The routine use of deception can result in poorly crafted and trivial experiments that are unjustifiable. This chapter examines the factors that may make deception research wrongful and harmful. It indicates ways to achieve valid research objectives without wronging or harming subjects and discusses the appropriate uses of dehoaxing and desensitizing.

7.1 WHY IS DECEPTION USED IN RESEARCH?

There are four defensible justifications for deception research. Deception may be the only viable way to accomplish the following research objectives:

1. To achieve stimulus control or random assignment of subjects.
2. To study responses to low-frequency events.
3. To obtain valid data without serious risk to subjects. For example, in research on conflict, one may employ an accomplice or confederates who will not escalate the conflict beyond the level needed for the purposes of the research.
4. To obtain information that would otherwise be unobtainable because of subjects' defensiveness, embarrassment, shame, or fear of reprisal.

An indefensible rationale for deception is to trick people into research participation that they would find unacceptable if they correctly understood it. *If it is to be acceptable at all, deception research should not involve people in ways that members of the subject population would find unacceptable.*

A flawed rationale is to promote spontaneous behavior in a laboratory setting. There is abundant evidence that participants in laboratory research, especially college students, typically assume that deception will occur and engage in deceptive hypothesis testing of their own (Geller, 1982; Orne, 1969). Contrary to what many researchers claim, subjects are respectful of consent procedures in which they are asked to permit the researcher to withhold some information until after they have participated, with a sincere promise of full debriefing afterward.

Case 7.1: Deceiving to Study Conformity. Solomon Asch (1956) studied conformity by telling subjects that they were participating in a perception experiment, in which each member of their group would select the line believed to be the same length as the standard line. Unknown to the subject, the other seven members of the group were confederates. On the first two trials, most of the seven confederates made the correct match. From the third trial on, the confederates all agreed on the wrong answer. Asch reported that the (real) subjects looked bewildered and anxious. Thirty-three percent of them gave the same wrong answer as the confederates, while the rest gave the correct answer despite obvious feelings of discomfort and confusion. The 33% error rate in the false majority condition was in sharp contrast to the 7% error rate in the condition without the confederates. Most subjects doubted their own judgment, and one-third of them caved in to the majority opinion. Following each subject's participation, a sensitive debriefing was conducted in which the procedure was fully explained and subjects were assured that their responses to conformity pressures were normal.

Asch's conformity study required deception for all four of the defensible reasons mentioned above. The first three rationales for Asch's deception—control and random assignment, study of a low-frequency event, and avoidance of real, and possibly dangerous, interpersonal conflict—obviously apply, but the reader might wonder whether the fourth rationale is pertinent: Are people that defensive about their conformity? Would self-report data have been as accurate? Apparently not. Wolosin, Sherman, and Mynatt (1972) found that individuals perceive most other people as conforming sheep, but report that they themselves are independent of group influence.

Did Asch's subjects see through the deception? Since the research was performed nearly a half-century ago, before deception became a method of choice among social psychologists, the subjects may actually have been naive and spontaneous.

7.2 ALTERNATIVES TO DECEPTION

The ethical and methodological problems of deception research have received much attention in recent decades (e.g., Geller, 1982; Kelman, 1967, 1972; Seeman, 1969). The emphasis in social research has shifted increasingly to field settings where deception is not tolerated (see, for example, Chapter 11, on community-based research). Also, the rise of IRBs has caused researchers to think twice about their methods, rather than regard deception as the unquestioned method of choice. Out of this confluence of events, three main alternatives have emerged.

1. Simulations—mock situations in which subjects are asked to act as if the situation were real—are effective ways of exploring social behavior. Three basic types of simulations have been developed: game simulations, field simulations, and role-playing simulations; each has taken an important place in research.

In game simulations, subjects take roles under a particular set of rules and maintain them until a desired outcome is reached. For example, mock trials have been convened, using real jury candidates, real judges, and real cases (Boruch, 1976). Such simulations are highly realistic.

Field simulations lack firm rules, use highly realistic staged settings, and encourage subjects to believe they are participating in a natural event. For example, the Stanford prison study (Zimbardo, Haney, Banks, & Jaffe, 1973) involved students who volunteered but did not know whether they would play prisoners or guards. The students were unexpectedly picked up in real Palo Alto police cars and booked at the police station before being reminded that this was the study for which they had volunteered. Field simulations, which may last for days, are so highly involving as to produce extreme behavioral and emotional responses; although avoiding deception, therefore, such simulations may well raise other kinds of ethical concerns.

Role playing is the type of simulation that is best suited for the kind of experimentation in which deception has typically been used. The role player knows of any illusions that are created in the setting and is asked to act spontaneously, as if the situation were real. Like a method actor, the role player attempts to do what one would actually do in a particular (contrived) situation. When realistic props are used, as in a deception experiment, the behavior of role-playing subjects is often indistinguishable from that of their counterparts in deception experiments (Geller, 1978).

2. Ethnographic or participant observation methods, coupled with self-report, are used increasingly to study real behavior. As described in Chapter 11, most community-based research involves the researcher deeply enough in field settings that it is feasible to validate self-report against observations of naturally occurring behavior. In such settings, the key to accuracy is not the technical cleverness of deception, but rapport, trust, and good ethnographic and interviewing skills.

3. Consent to concealment may be obtained. There is now ample evidence that most subjects will gladly participate in research with the understanding that some details must be withheld until after they have participated, and that a full debriefing will follow. We turn now to a fuller discussion of concealment versus deception strategies.

7.3 CONSENT TO CONCEAL VERSUS DECEPTION

Five kinds of deception or concealment may be used to achieve the objectives discussed in 7.1. Three approaches involve consent and concealment. Note that only two of the approaches listed below actually deny subjects their right of self-determination and involve deception.

1. *Informed consent to participate in one of various conditions.* The various conditions to which subjects may be assigned are clearly described to subjects ahead of time. For example, most studies employing a placebo use this consent approach. Subjects know that they cannot be told the particular condition to which they will be assigned, as this knowledge would affect their response. Complete debriefing is given afterward. Subjects who do not wish to participate under these conditions may decline to participate.

2. *Consent to deception.* Subjects are told that there may be misleading aspects of the study that will not be explained to them until after they have participated. A full debriefing is given as promised.

3. *Consent to waive the right to be informed.* Subjects waive the right to be informed and are not explicitly forewarned of the possibility of deception. They receive a full debriefing afterward.

4. *Consent and false informing.* Subjects consent to participate and are falsely informed about the nature of the research.

5. *No informing and no consent.* Subjects do not know that research is occurring. Subjects assume that they are just engaging in "real life." The setting may be contrived, or it may be a natural setting in all respects but one—it may contain a spy.

Most research requiring deception can be done about as well with one of the first three forms of deception as with the latter two. In the two cases that follow, consider how a researcher's decision to apply the first, rather than the fifth, form of deception might influence the consequences of a study of the behavioral effects of LSD:

> *(Hypothetical case).* Subjects are asked to participate in a study of the effects of LSD on specified behaviors. They are told they will receive either a salt tablet (a placebo) or an LSD tablet, but that they cannot be told ahead of time which they will receive. Subjects experience affective states and may incorrectly attribute the cause—hence the need for a placebo group. Before being released, subjects are monitored for a period of one week in a hospital setting, to ensure that the treatment effects, if any, are gone. All subjects are debriefed on the fifth day.

> *Case 7.2: CIA Research on LSD.* A true example of deception with no consent or debriefing is the following study, which was supported by the CIA: The investigators set up an elaborate laboratory in a brothel. As clients arrived, they were given drinks containing LSD. The men's behavior was then filmed from behind a two-way mirror. The subjects were never debriefed. One subject committed suicide while under the influence of LSD. The cause of his suicide and details of the research remained secret until many years later when the victim's family found clues that led to the truth (Goldman, Clark, & Marro, 1975).

These examples are informative. The first four forms of deception readily permit the research to be conducted in a setting where accurate measurements, proper controls, and all necessary safeguards are readily at hand. However, the form of deception in which subjects do not realize that they are participating in research is more likely to occur in field settings where assurances of privacy and confidentiality, adequate standby medical aid, and other features of scientifically and ethically sound research simply cannot be provided.

The fact that there is no informing and no consent in a field setting, however, does not necessarily mean that the research would be judged harmful or wrongful. Rather, it is how society evaluates the behavior studied that largely determines the attribution of harm or wrong, as the following example might suggest:

Case 7.3: Cookies and Kindness. Isen and Levin (1972) investigated the effects of a person's positive affective state on subsequent willingness to help others. Feeling good was induced in half the subjects by handing out free cookies to them while they were studying in the library. The other half did not receive cookies. Half the subjects in each group were then approached by another person and asked if they would volunteer to serve as confederates in a psychology experiment designed to study creativity in students at examination time, as compared to other times of the year. There were also told that the experimental procedure was something students would find helpful. The other half of the subjects were asked if they would volunteer to serve as confederates in a study requiring that they drop books, make noises, and so on while the experimenter unobtrusively recorded students' reactions to distraction at examination time, as compared to other times of the year. They were also told that the distraction would be generally regarded by students as an unpleasant annoyance. The results of this study indicate that students who received cookies were far more willing to engage in helpful behavior and less willing to engage in distracting behavior than those who did not receive cookies. A debriefing and discussion period followed each subject's reply. Subjects indicated that they had regarded the handing out of cookies as a kind gesture, unrelated to the subsequent request.

The deception employed in Case 7.3 mildly induced good public behavior. But what of field research that powerfully induces bad behavior? Who wants to be spied upon, especially when engaging in private behavior, and even more so when engaging in bad behavior? Going a step further, who wants to be tricked (induced) into performing private deeds or bad deeds for the express purpose of affording someone else the opportunity to spy and record those deeds? Such treatment, obviously, should not be foisted upon those who would object. But for the sake of social science, some hearty souls might willingly sacrifice a bit of their dignity and privacy, and consent to research procedures involving concealment and powerful induction of private behavior, especially if the research were of considerable social importance and were competently performed. At this point, *confidentiality* becomes a critical means of preventing harm. Whenever possible, data should be gathered anonymously. If unique identifiers must be gathered, however, all possible precautions must be taken to ensure confidentiality (see Chapter 6).

7.4 MINIMIZING WRONG AND HARM

In writing this chapter for a broad audience—including some who conduct deception research for a living and others who are dead set against it—I can only hope that I have brought all readers to recognize two things: (a) Some important forms of behavior vanish under obvious scrutiny; concealment or deception is sometimes necessary in research, and (b) the more objectionable forms of deception are unnecessary. If I have succeeded thus far, we can now list some questions one should ask before setting out to study one of these more elusive forms of behavior that call for deception or concealment.

1. Is participant observation, an interview, or a simulation method likely to produce valid and informative results?
2. Would one of the "consent to concealment" procedures work?
3. Is privacy invaded? If so, have subjects consented to something like this, and have members of the subject population who understand the procedure decided that they would be willing to participate in such a study? Is confidentiality or anonymity absolutely assured?
4. If studying bad behavior, have subjects consented to something like this, and have members of the subject population who understand the procedure decided that they would be willing to participate in such a study? Is confidentiality or anonymity assured?
5. If studying private or bad behavior, is it induced? How strongly?
6. Is debriefing impossible? Debriefing deception research includes dehoaxing (revealing the deception) and desensitizing (removing any undesirable emotional consequences of the research).
7. Is the study of such overriding importance and so well designed that deception is justified?

7.5 DEHOAXING

An obvious advantage of the first three concealment methods described in 7.4 is that they are easily dehoaxed. When the latter two methods are used (consent and false informing, or no consent and no informing), dehoaxing is not always easily accomplished. Having been deceived once, why should subjects believe the next thing they are told?

Whatever the device used to deceive, it should be demonstrated to be fraudulent. When technical deception (misrepresentation of objects or procedures) is employed, a convincing demonstration of the deception is usually easy to arrange. For example, when a subject is given false feedback, supposedly based on a test performance, the dehoaxing might include returning to subjects their own tests, still in sealed envelopes, just as the subjects turned them in (Holmes, 1976). When role deception is used, the dehoaxing should include an introduction of subjects to the real person, along with the presentation of whatever information is needed to establish the person's real identity. Implicit deception (in which subjects naturally generate wrong assumptions that the investigator is careful not to correct) is the most difficult to dehoax, since essentially subjects have misinformed themselves. Here, it is tactful and beneficial to assure subjects that anyone would misinterpret the situation and that the research was designed so that misinterpretation would occur.

After dehoaxing, the rest of the debriefing procedure, as described in Chapter 4 (4.6), should be carried out, including a discussion of the nature of the observation, the data collection, the design, and a fairly detailed explanation of the purpose of the research.

Generalized mistrust is psychologically harmful, and the social scientist is obligated to prevent its occurrence or to remove any generalized mistrust engendered by the research. Hence, the practice of double deception is particularly harmful:

> This practice involves a second deception presented as a part of what the participant thinks is the official post-investigation clarification procedure. Then some further measurement is made, usually using covert means to assess the impact of the conditions upon the true dependent variable. In such cases there is a particular danger that the participant, when finally provided with a full and accurate clarification, will remain unconvinced and possibly resentful. Here, confidence in the trustworthiness of psychologists has realistically been shaken. (American Psychological Association, 1973, p. 80)

However, dehoaxing does not necessarily return subjects to their prior emotional state, and may even cause emotional difficulties.

7.6 DESENSITIZING

The investigator should detect any undesirable emotional consequence of the research for participants and restore them to a frame of mind that is at least as positive and constructive as it was when they entered the study. Deception research may induce subjects to reveal information they would prefer to have kept secret, or it may inflict on them new insights about their own personal weaknesses. Embarrassment, self-doubt, guilt, fear of damage to one's reputation, and other negative emotional consequences may ensue. Desensitizing should alter subjects' feelings concerning the way they behaved or were treated in the study, so that they are restored to a state of emotional well-being. The investigator should have adequate facilities to handle any emotional reactions to stress that may arise from participation in the study or the debriefing. In addition to providing effective and caring desensitization immediately after the research session, the researcher should make other sources of counseling or psychotherapy available to the subjects.

As discussed at length by Holmes (1976), it is not clear that desensitizing efforts always succeed in removing all of the self-doubt engendered by some deception. Hence, it is important that studies not involve treatments that are damaging to self-esteem in the first place. In any event, whatever desensitizing procedures are used, they should include processes or material designed to address the following concerns:

Confidence in science. Subjects should receive adequate explanation so that they will consider the research reasonable and will feel no loss of confidence in either the investigator or science.

Risk control. Subjects should be informed of any risks the researcher has anticipated, and of the steps taken to minimize them (e.g., use of confidentiality-assuring procedures). Any unwarranted concerns subjects may have about harm that could result from their participation should be detected, fully discussed, and allayed. The researcher should be receptive to subjects' concerns about unanticipated forms of harm and should be prepared to do everything possible to prevent such harms from materializing.

Questions. Subjects should be given an opportunity to ask questions about the study, and to have their questions answered satisfactorily.

Withdrawal of data. Subjects should be given an opportunity to withdraw their data from the study. This ethical requirement follows from the principle of respect for persons and is codified in the American Psychological Association's Principle 5, which states:

> Ethical research practice requires the investigator to respect the individual's freedom to decline to participate in research or to discontinue participation at any time. The obligation to protect this freedom requires special vigilance when the investigator is in a position of power over the participant. The decision to limit this freedom increases the investigator's responsibility to protect the participant's dignity and welfare.

The American Psychological Association's Ad Hoc Committee on Ethical Standards in Psychological Research notes that if the investigator has withheld or distorted information that would influence the participant's consent to participate, the right to withdraw data substitutes for the right to refuse to consent to participate (1973).

The right to withdraw one's data is particularly important when data on private behavior are gathered, and subjects experience discomfort about having revealed things that they would prefer not to have revealed. Just as respondents to an in-depth interview may regret having been skillfully drawn into a discussion in which they revealed more personal data than they would have wanted to reveal, so may participants in deception research discover, upon being debriefed, that they have revealed something about themselves that they would not have chosen to reveal if fully informed. In either case, subjects should have the option of withdrawing their data.

7.7 WHEN NOT TO DEHOAX

When studying behavior that is socially perceived as negative and when the behavior is typical of the subject, desensitizing is usually unnecessary, and dehoaxing may even be harmful.

Example 1: As any parent or child psychologist knows, young children steal—often innocently and without a sense that stealing is wrong. Suppose a researcher studies stealing in young children by making toys or coins available for stealing. After observing the children stealing, should the researcher then debrief them, by telling them that they were observed stealing, or by telling them that the study was about stealing? The researcher's business is not to punish, to shame, or to try to produce accelerated moral development, but only to study how children behave. The ethical guidelines of the American Psychological Association hold that children should not be debriefed when debriefing would upset them.

This example raises another problem: In obtaining parental permission to perform the research, should the researcher agree to inform parents of their child's response? Not necessarily. If such an agreement is made, however, the researcher must also give parents information about *typical* behavior for children of that age, and about *appropriate* parental responses to stealing by children of that age. With this background, most parents would handle feedback about their own children responsibly. However, the researcher must be sensitive to the needs of particular parents and children, and seek consultation about the handling of difficult cases.

Example 2: Some adults have attitudes or habits that the researcher (and most of society) would deem undesirable, but which the adults in question view positively and would not readily change. After participating in a deception study of parental authoritarianism, for example, should subjects be told that their authoritarian child-rearing practices have been studied? In all cases, debriefing should be done without demeaning subjects.

RECOMMENDED READINGS

Sieber, J. E. (1982, November). Deception in social research I: Kinds of deception and the wrongs they may involve. *IRB: A Review of Human Subjects Research*, 1-2, 12.

Sieber, J. E. (1983, January). Deception in social research II: Factors influencing the magnitude of potential for harm or wrong. *IRB: A Review of Human Subjects Research*, 1-3, 12.

PART III

Risk/Benefit Assessment and Planning

Prior chapters have discussed issues of research risk and benefit. This section examines how risk and benefit are assessed.

WHY ASSESS RISK AND BENEFIT?

To minimize or avoid risk and to maximize the benefit that may result from research, one must first identify the kinds of risk and the kinds of benefit that are possible within a study. A discussion in the protocol of the findings of this assessment and the methods and procedures to be employed to maximize benefit and minimize risk are crucial to the IRB's evaluation of the research plan.

Some vitally important social research, such as research on runaways, child prostitutes, or drug dealers, may necessarily involve risk. Such research is acceptable if it is well designed, if it is conducted by a competent investigator, and if risk/benefit assessment and planning have occurred. Minimal harm, valuable knowledge, a publication, and enlightened public policy would be among the expected outcomes. A similar amount of risk would not be acceptable in an undergraduate project.

MISCONCEPTIONS ABOUT RISK/BENEFIT ASSESSMENT

The term *risk/benefit assessment* invites misconceptions. These should be corrected now:

1. A ratio is not actually computed; most risks and benefits of research cannot be quantified.

2. Some risks and benefits cannot be identified accurately before the research is performed.

3. It is impossible to consider all possible risks and benefits. There are many forms of good that people want to enjoy and many harms they wish to avoid. Forms of good (e.g., freedom, self-respect, friendship, contentment, health, self-expression, understanding, pleasure) and the harm implied by loss of each are merely abstractions of the many specific emotional, physical, social, legal, and economic forms of good and harm that could arise in connection with research. In assessing social sensitivity, risk, and benefit, the researcher seeks to focus upon the most important of these, but cannot possibly anticipate everything.

4. Risk and benefit cannot be identified for each subject individually. One subject's risk may be another's benefit.

WHAT IS RISK/BENEFIT ASSESSMENT?

Risk/benefit assessment weighs the risks, or costs, of the research to subjects and to society, against its benefits. These risks and costs include ones incurred when socially sensitive research offends some sector of society, resulting in backlash against the subjects, the researcher, the participating institutions, and the cause the researcher had hoped to promote. IRBs assess risk to subjects only, as opposed to risk to groups in society or risk of offending persons who happened to hear about the research.[1]

A risk/benefit assessment is based on most of the same forms of inquiry that should inform other aspects of research design, including common sense, a review of the literature, knowledge of research methodology, ethnographic knowledge of the subject population, perceptions of pilot subjects and gatekeepers, experience from serving as a pilot subject oneself, and input from researchers who have worked in similar research settings.

Two keys to risk/benefit assessment and planning are (a) knowing where to look for potential risks (costs, harms) and benefits, and (b) engaging a wide spectrum of players in the risk/benefit assessment to obtain diverse perspectives and value orientations. No single source can say what potential risks and benefits inhere in a particular study. Risk, and the ethical issues raised by risk, cannot be defined by a few specifiable dimensions. The benefit and justifiability of research depend on the whole nature of the research process and on the values of the persons who judge the research. Thus, a researcher, those with whom he or she consults, and the IRB can only estimate whether a proposed project would turn out well. However, the potential for risk and benefit can be assessed in advance, and throughout the research process:

Case 1: In consultation with the IRB, with colleagues in his field, and with a local community-based drug treatment program, an investigator began the design of survey research on black teenage crack users.[2] After some initial planning, he invited prospective subjects to focus group meetings. For the design of the interview, he needed to know their beliefs about crack, where they got it and how they used it, their knowledge and practice of safe sex, and their sexual behavior in general. He was not sure how to recruit subjects, establish rapport, or phrase the interview questions, but he knew he could relate effectively to focus group members; he is a black man who easily engages youth in conversation.

The focus groups met regularly over lunch. Drug abusers often neglect nutrition, but when given food, may be ravenously hungry. The investigator learned their concerns about their lives and mentioned ways he would like to help them. He asked their advice on how to organize the interviews, what to ask, and what fears respondents would have about participating candidly. He asked what they, and the actual subjects, would like to have in return for their efforts. The investigator found that many focus group participants wanted to learn to be interviewers and some wanted drug treatment. These participants helped develop the interview questions, the recruitment strategy, and strategies for providing desired services to participants and interviewers. Those who were trained to become interviewers had access to crack houses where the investigator could not go. As community members, they could get candid responses to interview questions. As interviewing ensued, supervision revealed that an interviewer had regaled his sister with details of an interview in which a man they knew had named all the women in the neighborhood with whom (he said) he had slept. More discussions of confidentiality were held with the assistants, and the interviews were resumed. This research culminated in the publication of useful information about the relationship of crack use to high risk sexual behaviors. The results have been useful to drug treatment programs.

HOW DOES AN INVESTIGATOR IDENTIFY RISKS AND PLAN BENEFITS?

Throughout Part III and other parts of this book, issues of social sensitivity are raised. The federal regulations governing IRBs are not concerned with social sensitivity, but only with harm to human subjects. However, the investigator who ignores the sensitivities of gatekeepers, community members, and society at large risks having the research prevented, interrupted, or maligned by powerful forces other than the IRB.

NOTES

1. CFR 46.111 states: The IRB should not consider the possible long-range effects of applying knowledge gained in the research as among those research risks that fall within the purview of its responsibility. However, no researcher would want to invite the destruction that can result from offending social sensitivities.

2. Under 46.408 of the Federal Regulations governing human research, one need not obtain permission for research on minors, provided parents or guardians are unreachable, or unlikely to act in the best interests of their children. (See Chapter 10.) Young crack users and dealers often live apart from any parents or guardians, who would likely be unreachable in any case. IRB consent was given to conduct the study without seeking parental approval. However, some laws in some states are more restrictive.

8

Recognizing Elements of Risk

All ethical codes governing human research require that investigators identify possible risks ahead of time and plan their research so that subjects' rights are protected. For example, the Code of Ethics of the American Psychological Association (Principle 9G) states that:

> The investigator protects the participant from physical and mental discomfort, harm, and danger that may arise from research procedures. If risks of such consequences exist, the investigator informs the participant of that fact. Research procedures likely to cause serious or lasting harm to a participant are not used unless the failure to use these procedures might expose the participant to risk of greater harm or unless the research has great potential benefit and fully informed and voluntary consent is obtained from each participant. The participant should be informed of procedures for contacting the investigator within a reasonable time period following participation should stress, potential harm, or related questions or concerns arise. (APA, 1982, p. 6.)

Few things concern an IRB more than an investigator who blithely states that no risk is involved in proposed research, when risk is evident to the IRB. Such a naive assurance suggests that the investigator is insensitive and likely to cause needless harm and upset.

One reason some researchers make no attempt at risk assessment is that the task seems impossible. It is indeed impossible to identify all relevant risk, but it is important to consider possible risks and be prepared for dialogue with the IRB. Because of the complexity of risk assessment, no two people will arrive at exactly the same conclusions. Unfortunately, investigators often become angry and defensive when the IRB raises the possibility of risks that the investigator has not considered. One purpose of this chapter is to provide a framework for identifying possible risks, thus creating a basis for dialogue, rather than defensiveness.

8.1 WHAT IS RESEARCH RISK?

Risk refers to the possibility that some harm, loss, or damage may occur. There are various kinds of risk in research, including the following:

- *mere inconvenience*, such as boredom, frustration, and taking up time that the subject might more profitably spend otherwise.
- *physical risk*, such as the possibility that one will get a black eye from participation in an exercise physiology experiment.
- *psychological risk*, such as the possibility of getting a black eye and becoming depressed about attending events wherein wrong inferences may be drawn about the cause of the black eye.
- *social risk*, being rejected for having that black eye.
- *economic risk*, being passed over for employment in favor of another candidate who interviewed without a black eye.
- *legal risk*, being arrested and interrogated about the possible connection between the black eye and a brutal assault that left a neighbor comatose.

These examples illustrate risks to subjects. As we shall see, however, these kinds of risks may apply to others as well, including the researcher and other persons associated with the research, the setting where the research is performed, and other settings and populations that are somehow affected by the research.

8.2 WHY RESEARCH RISK IS DIFFICULT TO ASSESS

Risk assessment is not intuitively easy. Most investigators are sensitive only to the risks they have already encountered and may fail to assess major risks in new research settings. Even a chapter such as this cannot point to all possible research risks. The goal of this chapter is to present a model that enables one to recognize circumstances likely to produce risk.

The model can be illustrated as shown in Figure 8.1, whose three dimensions consist of the following: (a) four aspects of scientific activity, (b) eight risk-related issues, and (c) seven general kinds of vulnerable entities. A quick mental calculation shows why each kind of risk is not illustrated here by a case study. The $4 \times 8 \times 7$ matrix yields

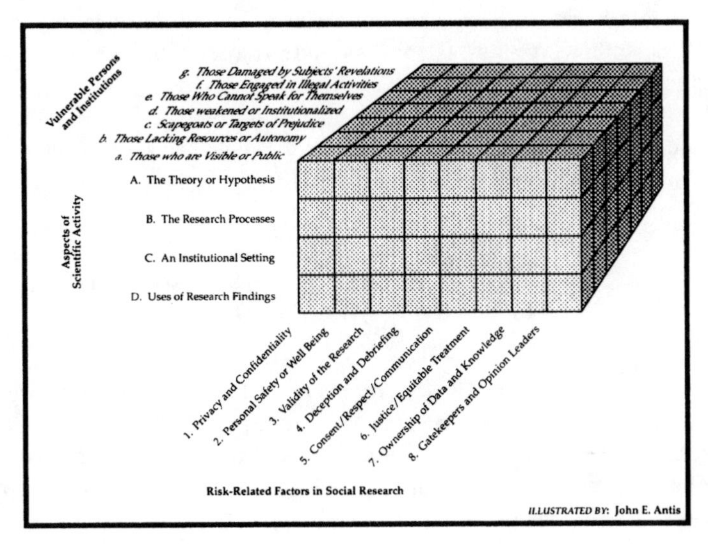

Figure 8.1.

224 cells, each containing the elements of a distinctly different case scenario.

The model should be read as follows, taking as an example a risk occurring at the intersection of *C, 1, d*. Because of an *aspect* of scientific activity (e.g., its setting—university administrative offices), a *risk* arises (involving, e.g., confidentiality—the interviewers are members of the university community), and this risk is serious because the *subjects* are weak members of the institutional setting (i.e., nonmanagerial employees who are being interviewed about their supervisors). Let us examine one more cell intersection:

Cell B, 6, b: A research process is unjust to nonautonomous persons who cannot complain effectively. Hypothetical example: In a program of applied research on math learning, children are randomly assigned to various 4-month cable-TV viewing conditions. Condition A is far superior to the standard curriculum, while Condition B has a damaging effect on math learning. After the 4-month program, all students go on with the standard curriculum. (Effects should be monitored, and all subjects should be provided with the superior or desired program after the data have been gathered. In treatment programs that could cause irreversible harm, subjects should be closely monitored and switched to the desirable program as soon as undesirable effects are observed [Conner, 1982].)

The reader is urged to refer to the matrix in Figure 8.1 while reading sections 8.3 through 8.5, which describe the matrix and illustrate ways in which risk may arise.

8.3 STAGES OR ASPECTS OF SCIENTIFIC ACTIVITY THAT MAY INVOLVE RISK

Four aspects of scientific activity may involve risk: (1) the theory or research idea, (2) the actual research process, (3) the setting of the research, and (4) the dissemination or uses of the research. The federal regulations are concerned only with 2 and 3, but any conscientious researcher will consider 1 and 4 as well.

8.3.1 The Theory or Research Idea

A new idea may change cultural values in fundamental ways, as occurred in response to the theories of Freud, Darwin, and Copernicus. Examples are plentiful. To avoid confusing the effects of ideas with the effects of research results, consider an idea of Freud's that was unsupported by research:

> Many of Freud's adult clients told him that they had been sexually abused by their parents. Freud announced this finding to his colleagues, who expressed shock and dismay that he would believe this of fine Christian parents. In response to his critics, Freud theorized that children's accounts of being sexually abused are products of fantasy.
>
> It is now recognized that approximately one child in three has experienced sexual abuse, either at the hands of adults or of older children, and that young children are incapable of describing sexual abuse based on fantasy alone. Moreover, until a decade ago, claims of sexual abuse in childhood were not taken seriously, largely because of Freud's ideas.

One might suppose that Freud's theory was so damaging because it was untested. This is not necessarily so; scientists often find what they seek, even when it is not there (Mitroff & Kilmann, 1979). But even if science were always self-correcting, research can still cause damaging misinformation. Consider a hypothetical case involving cells *A, C, D, 5, b, d*:

> A researcher is personally persuaded that boys are innately inferior to girls in social skills. In the course of obtaining informed consent, he describes his hypotheses to principals, teachers, and parents so persuasively that they become sensitized to evidence of poor social skills in boys. Consequently, some fail to challenge boys to develop social skills. Others decide that it is best to cultivate "the brutishness of the male." Will the research results correct these misconceptions? Probably not. Most researchers forget to give gatekeepers and subjects feedback after the data have been analyzed.

Four ways to reduce the risk of false confirmation or dissemination of damaging ideas are available to the researcher: (a) Recognize that the null hypothesis may very well be true; (b) design the research so that each theoretical orientation is tested fairly (Mitroff & Kilmann, 1979), or identify and develop competing hypotheses, consulting or collaborating with scientists who support alternative hypotheses; (c) remember the limitations of the models and measures employed, always announcing up front and reminding yourself and others that application and generalization to other populations must be done with caution; (d) upon completion of the research, share the documented data with other scientists who want to verify the findings or test alternative hypotheses with the data (Fienberg, Martin, & Straf, 1985).

8.3.2 The Research Process

The research process refers to steps involved in the actual conduct of empirical study. These steps typically include designing the research, recruiting the subjects, obtaining informed consent, administering the treatments (tests, interviews, and so on), gathering the data, and analyzing and interpreting the results. The following examples illustrate issues arising in the research process:

Recruitment. Inner-city men who have tested positive for sexually transmitted diseases are recruited for a study of safe sex practices. Community members figure out why people are being recruited, and the grapevine goes to work. Those recruited become stigmatized in their community; they do not return for interviews (Case, personal communication).

Induction. Subjects are invited to appear for on-campus job interviews. In fact, a deceptive study of self-presentation is being conducted. A student with financial problems has his suit dry cleaned, buys new shoes, and takes a day off from work to attend the interview, hoping to get a better-paying job so that he can complete school. Upon debriefing, he becomes so angry at the investigator that he vows to find and tell every prospective subject about the deception.

Consent and experimental treatment. Subjects in a study of correlates of drug abuse do not like the questions they are being asked. The questions seem different from what they agreed to. Because the research design was not adequately explained to them, they do not understand why some of their friends are asked different questions. They are suspicious about how their data might be used against them, and decide to quit the outpatient drug recovery program that sponsored the research.

Data gathering. Research assistants know the subjects and gossip about confidential information.

Debriefing. The researcher explains to parents who have just been studied in interaction with their young children that the research is designed to demonstrate how certain patterns of interaction produce unruly youngsters. As most children spend some time being unruly, many of the parents go away blaming themselves, their spouses, or other household members.

Data analysis. When cell size is too small to yield meaningful information, the researcher nevertheless fails to collapse cells and reports unreliable conclusions. Data summaries are published in which some cells contain so few cases that people familiar with the setting can deduce the identity of certain anonymous respondents, who have reported engaging in undesirable behavior.

Data sharing. The raw data are archived, without removal of unique identifiers, at the university library, where they are available to students and faculty. A blackmailer puts some of the juicier personal data to lucrative use.

These are only a few of the many kinds of risk that may arise because of the research process. Other kinds of risk arising within the research process are discussed elsewhere in this book.

8.3.3 The Institutional Setting of the Research

One readily thinks of research as a matter between researcher and subject. However, there usually is a "third party"—the setting or organization in which the research occurs. Aspects of the research that would otherwise be harmless may be risky because of the nature of the setting. The setting may be a community, a workplace, a hospital or clinic, a prison, a school, a church, a service organization, a professional organization, a recreational setting—any setting that has some kind of structure, culture, or interests of its own.

In most cases, settings have gatekeepers and impose rules on those who want to do research therein. When the research is done in one's own institution, the gatekeepers may be the subject pool coordinator, the IRB, the office of sponsored projects, and so on. In field research, gatekeepers are representatives of the setting the researcher wants to enter, for example, a school principal, a retirement home director, or a recovered drug addict who works as an advocate and community outreach person to street addicts whom the researcher hopes to interview.

Gatekeepers have the power to help researchers understand and establish rapport with the research population. They also have the power to negotiate conditions that are acceptable to those they serve. Thus, the researcher may expect to have to change some details of the research plans to suit the priorities of the setting and to contribute positively to that setting. Be aware, however, that gatekeepers and those they serve are not always interested in objectivity. They would not want the researcher to discover something that would be damaging to them or to their organization. They may even pressure the researcher to produce results that make them look good; hence, the researcher must be careful not to enter into unethical agreements with gatekeepers. In any case, gatekeepers will want the researcher to leave the setting, its staff, and its clientele in a positive state.

The following are two examples of risk and harm by an outside researcher who failed to understand or care sufficiently about the culture and interests of the setting:

> Organizational research is often conducted in a corporation by an academic social scientist who is more attuned to the norms of academia than to those of the corporation. Because both the investigator and the corporate officers have considerable autonomy, there is much potential for conflict and harm. The organization may make personnel decisions based on data that the investigator obtained from employees with a well-intended promise of confidentiality given by the investigator. Later, the investigator may release news about the research findings to a reporter. The news publication may harm the organization's standing with its customers, stockholders, personnel, competitors, or the government. The officers of the organization may halt a long-term research program in midstream and thus harm both science and the career of the scientist. (Based on an actual case described by Mirvis, 1982)

> Because medical and psychiatric clinics have their own concerns about maintaining their services uninterrupted and protecting confidentiality, they provide entrée to outside researchers only after a lengthy negotiation and approval process (Sieber & Sorensen, 1992). The following is a hypothetical case of the kind that clinic directors hope to avoid: Given entrée to a clinic that serves pregnant teenagers, researchers conduct a repeated measures study of the development of attitudes toward maternal and infant health care. In violation of their agreement with the clinic, the researchers add sensitive personal questions that the subjects find highly objectionable. Not knowing that the researchers are solely to blame, the subjects fear to complain, lest their medical care be curtailed. Some solve the problem by giving dishonest responses to those questions. Some drop out of the study. Some drop out of the clinic program. Finally a particularly assertive teenager tells the clinic director what she thinks of the research. The researchers are asked to leave.

In other cases, the researcher may be an insider who is more powerful than any gatekeepers or there may be no effective gatekeepers or subject-advocates. Thus, mental hospitals have, in the past, experimented with behavior modification approaches designed to make patients easier to handle; students may be required to participate in a subject pool with no real opportunity to decline; men and women serving in the military may be required to serve as subjects in sleep deprivation research; and courts of law may seek to study the effects of various sentencing, parole, and treatment conditions on juvenile delinquents who are randomly assigned to appropriate but different conditions.

Because individuals vary in their degree of personal autonomy, and institutions in their degree of control and coercive power, the kinds of harm that may result from research in institutional settings range from the killing of prisoners in Nazi concentration camps and the imprisonment of scientists as war criminals, to the inconveniencing of college sophomores and tarnishing of reputations of faculty members. As these examples suggest, the degree of harm that can occur because of the coercive power of the institution is affected by the nature of the research process itself, and can be further compounded if the research idea is intrinsically insulting or damaging to persons.

8.3.4 Uses of the Research Findings

If mere ideas can influence world events, "proven" ones can be even more influential. In our enthusiasm to use what we have learned from research, it is easy to forget that many findings are based on measures of dubious reliability, or account for so little of the variance in these measures that they should not be considered practically useful. Findings based on one population or setting may not generalize to other populations or settings. Findings that are useful for one purpose may be misused in other instances. The social sciences deal with ideas that readily appeal to our personal values or politics. Scientists themselves are often incapable of separating their personal values and perceptions from objective observation (Doris, 1982). A good example is found in the early twentieth century, when public policy was adopted on the basis of the various ideas that scientists promoted regarding the mentally retarded. As a result of that public policy, the mentally retarded were treated inhumanly; vestiges of that public policy linger today. The case is old, but the lessons to be learned remain current.

> Around the beginning of the twentieth century, many social scientists considered mental retardation the most serious problem in the United States (Davies, 1930). They warned policymakers that unless the retarded were prevented from reproducing, mental

retardation would assume greater and greater proportions with each new generation, leading to more crime, alcoholism, prostitution, poverty, and ultimately, the extinction of Western civilization.

Sarason and Doris (1969) have shown that this concern about mental retardation is attributable to two advances in psychology: the rediscovery and extension by psychologists of Mendel's theory of inheritance, and Binet's development of a measure of children's intelligence. Both Goddard at the Vineland Training School and Kuhlman at the Minnesota Institution for the Feebleminded showed how quickly and conveniently the IQ test could be used to classify the mentally retarded. Goddard's study of the Kallikak family in 1912, coupled with a reanalysis of his data by the great Harvard geneticist, E. M. East, was put forward as proof that feeblemindedness is hereditary and is transmitted in accordance with the Mendelian model of recessive traits—a conclusion that is now thoroughly discredited.

Goddard then administered IQ tests in English to non-English-speaking immigrants passing through Ellis Island. Based on their performance, Goddard claimed to have discovered that emigrating Jews, Italians, Russians, and Poles would contribute to the intellectual deterioration of the American populace. During the 1920s Congress enacted the most restrictive immigration laws this country has seen. Noted psychologists, geneticists, sociologists, and psychiatrists advocated institutionalization and involuntary sterilization to remove the mentally retarded from society and prevent their reproduction. From 1910 to 1923, institutionalization rates for retardates more than doubled, and between 1907 and 1926, 26 states passed sterilization laws (Doris, 1982).

The combined power of the ideas and the public policies that came into existence at the beginning of this century made it difficult for valid opposing ideas to gain visibility. Doris (1982, p. 205) comments that Goddard and others:

> [H]ad created in the early decades of this century such a climate in the scientific and professional communities that Walter Fernald (1919), director of the Waverly Institution for the Feebleminded, withheld for two years research data demonstrating that former inmates of his institution—who for the most part had left without official blessing—had, in fact, adjusted quite well to the social and economic demands of the outside world. His defense was that the data were so contrary to the accepted theory for the management of the feebleminded that he hesitated to make them public. (Davies, 1930)

By the 1930s social attitudes were changing. The Great Depression caused people to realize that economic failure and unemployment were not necessarily because of bad genes. In addition to the new social zeitgeist, new data and better application of theory also caused social scientists to question earlier ideas. Attitudes toward the retarded began to change. Due to financial limitations, institutionalization had not occurred to the degree advocated, and squeamish local authorities had not enforced sterilization laws strictly. However, special education classes for the retarded continued to grow (Grant & Eiden, 1980). Although leading social scientists advocated segregated special education classes, their advocacy was not empirically based. There was no attempt to examine rigorously the relative merits of special education versus regular class placement (Sarason & Doris, 1979).

Within the past two decades, a new wave of scientists has supported the deinstitutionalization and mainstreaming of the mentally retarded. No research has ever conclusively supported claims either for institutionalization or for deinstitutionalization. However, a lack of conclusive research evidence is not always enough to keep scientists from speaking authoritatively when advocating ideas that are in keeping with their own values and that society is ready to hear.

New issues of equal or greater sensitivity continue to arise. A decade ago some research suggested a greater than chance association between the XYY chromosomes and criminal behavior. This finding led some politicians to propose that XYY-type children be identified and placed in a special treatment environment designed to prevent them from becoming criminals (Hook, 1973; Wiener & Sutherland, 1968). Today it is proposed by some that all members of our society be tested for AIDS antibodies. Note that universal screening for AIDS antibodies and XYY chromosomes, were they to occur, would involve a concatenation of problems: Imperfect research designs and inaccurate measurement methods would be used to identify persons who would then become objects of prejudice; powerful administering and enforcing institutions, social and economic sanctions, and the possibility of quarantine laws could be ruinous to those who test positive.

8.4 RISK-RELATED FACTORS IN SOCIAL RESEARCH

Eight key risk-related issues may arise in each of the above four stages of the scientific process and may affect any persons, institutions, or populations connected with the research. Through responsible planning, any of these issues might be transformed into a benefit. For example, a frail elderly woman might find it a frightening invasion of her privacy to be asked many personal questions by an unsympathetic interviewer from another ethnic group, yet she might be flattered and delighted by the attention of an attractive young researcher of her own culture, who also speaks her dialect (e.g., Loo, 1982).

8.4.1 Privacy and Confidentiality

A theory may invade privacy and diminish autonomy of persons by causing others to perceive them negatively. In the research process, a skilled investigator may extract more information than the subject intended to give. This invasion of privacy may be easier for investigators to accomplish if the research is done in an institution where the subject suspects that the information is available anyhow, and that lying might be punished. If the research is translated into public policy, prying into millions of people's lives (e.g., AIDS antibody testing and quarantine) could be legally mandated.

The ways in which issues of confidentiality may arise at each aspect of scientific activity are analogous to those of privacy. Fundamental to ethical research is an agreement between subject and researcher about how the data generated by the research will be controlled. See Chapter 6 for methods of assuring confidentiality.

8.4.2 Personal Safety and Well-Being

This is the least subtle of risk-related issues, and is included here for the sake of completeness. Might the theory reduce individual's safety, as in Freud's theory that children fantasize sexual abuse by parents? Does the research procedure contain elements that may cause direct physical or emotional harm (e.g., a torn muscle, extreme stress, damaged clothing)? Is the research setting unsafe (e.g., in a physically dangerous or frightening place)? Does the use of the findings cause physical harm?

8.4.3 Lack of Validity

Invalid research is bad science. It produces wrong information, the application of which may cause harm.

8.4.4 Deception and Debriefing

The very idea of deception in research damages the reputation of social science and may produce disingenuous responses from suspicious subjects, even in studies where deception is not employed. Some subjects may see through the deception all along, while others may be so deeply convinced by an induction (e.g., one that makes them feel inadequate) that no amount of debriefing removes the induction effects. In an institutional setting, the use of deception may destroy trust in the institution. The dissemination of socially sensitive studies involving deception may create considerable public backlash against social science and the funding of research. See Chapter 7 for solutions to these problems.

8.4.5 Informed Consent, Respectful Communication, and Consultation

What is communicated and not communicated by the researcher at each stage of the research process is critical. In the introduction to Part III, Case 1 provides an excellent example of effective communication (via focus groups) that quickly reached the entire research population through the community grapevine. Because the informed consent statement merely reinforced what people already knew, it was readily understood and believed. This communication process shaped the research idea, the research procedure, the relationship of the research to the community context, and the application of results within the community.

8.4.6 Justice and Equitable Treatment

Issues of procedural and distributive justice may arise at any stage of the research process. An idea that prejudices us against some sector of society is unfair, as is an experimental treatment in which resources known to be vital to subjects' well-being are withheld from subjects in one group and given to subjects in another. Such problems are compounded when they occur within an organization. If a vitally important treatment must be withheld from control group members during an experiment, it should be offered to them as soon thereafter as possible. Injustice may also occur in a variety of ways at the dissemination end of the process, as when the researcher takes no steps to encourage the dissemination of valuable knowledge to those who stand in greatest need. An excellent example of justice in dissemination is the fieldwork of Pelto (1988) on infant nutrition in various developing countries: Pelto routinely shares his data with appropriate governmental representatives so that the information can be used for the countries' own public policy purposes. He also provides training and consultation in data analysis, all as a way of repaying the host culture for its cooperation.

8.4.7 Ownership of Data and Knowledge

Ownership of data and knowledge refers to a complex of concerns about openness and democratic use of science that may arise at any point in the scientific process. Openness is vital to science. However, in some research for hire, the sponsor claims exclusive ownership of the data and the right to edit the results. The discerning reader may note a possible conflict between the principle of openness, or scientific freedom, and the need of applied researchers to accommodate the interests of gatekeepers in community settings where the findings may have political consequences. The researcher should gain all possible awareness of these concerns ahead of time in order to decide, in consultation with the gatekeepers, how they may be addressed without violating principles of openness or honesty. Often the problem is not what results are reported, but how they are framed.

Another possible conflict between openness and confidentiality arises when scientific data are shared. Most major funding agencies now require that researchers make their documented raw data available to other scientists, via either public archive or private arrangement. The legal rights of property ownership (e.g., to sell for profit, to withhold, to alter, to destroy) do not necessarily pertain to data that form the basis of publication, as these data are the basis of the researcher's claim to the validity of findings. Issues of data ownership and sharing are discussed in Sieber (1989) and touched upon in Chapter 6 (6.8).

8.4.8 Gatekeepers and Opinion Leaders

Most research is influenced in various ways by gatekeepers and opinion leaders. *Gatekeepers* function at each stage of the scientific process and include those who provide access to a research site, colleagues who will decide on one's promotion, IRBs, funders, and journal editors, to name but a few. *Opinion leaders* include those who create and support prevailing attitudes about scientific activity, who decide which theories are important and which research methods are "scientific," who are likely to object to socially sensitive research, or who would harass a scientist for his or her views or methods of research. Some specific opinion leaders include the press, advocates for one or another cause, politicians, senior scientists, religious leaders, editors, and science gadflies.

As an investigator, one must be sensitive to the views and influences of gatekeepers, yet one is also obligated to question and decide for oneself what is valid and ethical. For example, useless research efforts may result from employing methods that currently have high status in science, but are not appropriate or ethical for a given problem.

The scientist who ignores gatekeepers and opinion leaders runs three serious risks: (a) being unduly influenced by unrecognized forces in the research environment, (b) failing to create an explicit and constructive relationship with those who might have valuable input, and (c) being unprepared for possible criticism by gatekeepers and opinion leaders of a different persuasion.

8.5 PERSONS OR INSTITUTIONS THAT MAY BE VULNERABLE

Depending on circumstances, any person or institution may be vulnerable to research risk. Recognition of these vulnerabilities is vital to effective risk assessment. This chapter presents seven main kinds of vulnerability and approximately seven kinds of people within each. This amounts to about 50 categories of vulnerable people; obviously there are more.

Some are vulnerable because they are visible, famous, have deep pockets (i.e., are targets for lawsuits claiming damages), or may not have public sympathy. This category includes the investigator (who is visible), the research institution and other organizations (which have deep pockets), identified and well-known members of the population sampled in the research (whose reputation and opportunities may be altered by the generalizations about them that ensue from the research), well-known persons closely associated with the research participants about whom private information may be revealed, as well as certain kinds of research participants: the rich and the well-known.

Some are vulnerable because they lack resources or autonomy. This category includes the aged, minors, students, the poor, the uneducated, the mentally ill, the retarded, and the homeless.

Some are vulnerable because they are stigmatized. This category includes people with AIDS, with prison records, and with sexually transmitted diseases: minorities, homosexuals, women, rape victims, and the unemployed.

Some are vulnerable because they are in a weakened position, and perhaps in an institutional setting. This category includes alcoholics, prisoners, the hospitalized, military personnel, school children, employees, psychotherapy patients, and persons in crisis.

Some are vulnerable because they cannot speak for themselves. This category includes the dying, fetuses, and brain dead persons. It also includes persons who may be unable to assert their rights effectively: including illiterates, non-Western peoples, non-English-speaking people, primitive people, infants, and low-status people.

Some are vulnerable because their illegal activities may become known to law enforcement authorities through the research. This category includes drug abusers, juvenile delinquents, child molesters, wife beaters, illegal aliens, tax evaders, and political activists.

Some are vulnerable because they are associated with those who are studied, and may be damaged by information revealed about them. This category includes community members, family members, group therapy members, and co-workers of the research participants.

8.6 RESEARCHERS' PERCEPTION OF RISK

Having just reviewed the parts of the matrix in Figure 8.1, it is easy to understand that researchers who are narrowly focused on their research aims may readily lose perspective on risk factors. Researchers must balance complex roles, as scientists, individuals, and members of society, in which they are

> [S]ubject to the same societal processes as other members of society—the same economic pressures, the same social motivations and prejudices. A scientist may be a bigot; be personally exploitative of family, friends, and co-workers; be a supernationalist patriot; believe in war as a solution to international political conflict; or, be an anarchist, pacifist, militant radical, liberal dissenter, or outstanding altruist. These attitudes and behavior patterns may play some

> role in the choosing of a research problem or theoretical position. The value systems and attitudes of the individual may be more significant in the choice of colleagues, students and assistants. They are likely to be most relevant to the uses to which the scientist is willing to put the results of the work done. (Tobach, Gianutsos, Topoff, & Gross, 1974, p. 16)

Researchers' conceptualization, execution, and dissemination of research are not independent of the institutions or social settings in which it occurs (Kevles, 1970). Even apart from researchers' own values, society dictates what kind of research is supported, where it takes place, and the uses to which its results are applied. In the midst of all these influences and distortions, the scientist must seek an objective view of risk factors. IRBs look for evidence that investigators are:

- aware of possible sources of risk and have employed appropriate approaches to reducing those risks,
- in consultation with those who can help them understand and reduce risk,
- aware of their own biases and of alternative points of view,
- aware of the assumptions underlying their theories and methods and the limitations of their findings, and
- aware of how the media and other opinion leaders may translate their statements into flashy and dangerous generalizations.

9

Maximizing Benefit

When researchers vaguely promise benefit to science and society, they approach being silly. These are the least probable of good outcomes. Researchers typically overlook the more likely and more immediate benefits, which are the precursors of social and scientific benefit. Some of the most immediate benefits are those to subjects and—in the case of some community-based research—to their communities. These are not only easy to bring about, but are also owed and may facilitate future research access to that population. The *intermediate benefits*—to the researcher, the research institution and the funder, if any—are ones that any talented investigator with an ongoing research program can produce in some measure. It is upon these immediate and intermediate goals or benefits that any ultimate scientific and social benefits are likely to be based.

Case 9.1: A (Semi-Fictitious) Tale of Two Researchers. Psychologist A and Psychologist B each started an externally funded school-based experiment with instructional methods designed to improve the performance of students identified as learning disabled. Each sought to develop diagnostic and teaching procedures that could ultimately be used by school personnel.

Psychologist A began by discussing her intervention and research plans with school administrators, teachers, parents, and students and asking them to describe problems with which they would like to have assistance. Where feasible, she made slight alterations in her program to accommodate their wishes. She integrated the research program with a graduate course so that her students received extensive training in the conduct of research in the school setting, under her rigorous supervision. Part of the graduate students' activities involved video documentation of the steps taken to establish this research in the school setting, with appropriate concern for issues of confidentiality. This material was incorporated into her department's research curriculum.

Psychologist A provided the school faculty with various published materials on learning disabilities. She gave bag-lunch workshops and presentations on her project for school personnel. Some teachers became interested in trying her approaches in their classrooms. She showed them how to implement the approaches she was using in her research. Respecting their special knowledge of their students and subject matter, she urged these teachers to adapt and modify her approaches, as they deemed appropriate, and asked that they let her know the outcome. Together, they pilot tested adaptations of the methods concurrently with the formal experiments. All aspects of the project were videotaped. The tapes were edited into usable materials for teaching university courses in research methods, and for teaching teachers to work with learning disabled (LD) children.

Each set of LD children who participated in the various one-month training programs received special recognition for willingness to participate, and each received both the experimental training and training in how to assist other students with similar problems. At the end of each one-month session, each child received a certificate, a workbook on how to use the techniques on future lessons, and a manual on teaching the techniques to other students. These children also received opportunities to work with younger students, using the methods they had learned.

Two newspaper articles about the program brought highly favorable publicity to the researcher, the school, and the researcher's university. This recognition further increased the already high morale of students, teachers, and the researcher.

Unfortunately, of the six procedures employed, only two showed significant long-term gains after 6 months on standardized tests of learning and problem solving. However, the teachers who had been involved with pilot testing of variations on the treatments were highly enthusiastic about the success of these variations. When renewal of funding was sought, the funder was dissatisfied with the formal findings, but highly impressed that the school district and the university, together, had offered to provide in-kind matching funds. The school administrators wrote a glowing testimony to the promise of both the new pilot procedures and the overall approach. The funder supported the project for a second year. The results of the second year, based on modified procedures, were much stronger. Given the structure that had been created, it was easy to document the entire procedure on videotape and disseminate it widely. The funder provided seed money to permit the researcher, her graduate students, and the teachers who had collaborated on pilot testing to start a national-level traveling workshop, which quickly became self-supporting. This additional support provided summer salary to Psychologist A, to teachers, and to graduate students for several years.

Researcher B made no attempt to benefit others, except by hoping that her experimental procedures would be successful. Her initial treatment effects were even weaker than those of Researcher A, because of the difficulties of operating in a setting where she interfered with ongoing activities. The first year of her project was also the last. There were no publishable results.

This tale of providing benefits to the many stakeholders in the research process is not strictly relevant to all research. Not every researcher does field research designed to benefit a community. In some settings, too much missionary zeal to include others in "helping" may expose some subjects to serious risk, such as breach of confidentiality. Also, not all research is funded or involves student assistants. For example, many researchers engage in simple, unfunded, unassisted one-time laboratory studies to test theory. Even in such uncomplicated research, however, any benefit to the institution (e.g., a Science Day research demonstration) may favorably influence the institution to provide resources for future research; and efforts to benefit subjects will be repaid with their cooperation and respect.

9.1 THE RISK/BENEFIT REQUIREMENT

The ethical principle of *beneficence* requires a favorable balance of benefits to risks. The Federal Regulations 45 CFR 46.111(a) contain the following statement: "Risks to subjects (must be) reasonable in relation to anticipated benefits, if any, to subjects, and the importance of the knowledge that may reasonably be expected to result." In balancing risk and benefit, it is important to take those steps that make real benefit a distinct possibility.

	Subjects	*Community*	*Researcher*	*Institution*	*Funder*	*Science*	*Society*
Relationships	Respect of Researcher	Ties to University	Future Access to Community Site	Improved Town-Gown Relationships	Ties With a Successful Project	Ideas Shared With Other Scientists	Access to Trained LD Teachers
Knowledge/ Education	Informative Debriefing	Understanding of Learning Disability (LD)	Understand Research on LD	Improved Graduate Research	Outstanding Final Report	National LD Symposium	Media LD Presentations
Material Resources	Workbook	Books	Grant Support	Videotapes	Instructional Materials	Refereed Publications	Popular LD Literature
Training Opportunity	Tutoring Skills	Teacher Training	Greater Research Expertise	Graduate Student Training	Model Project for Grant Applications	Workshop at National Meetings	LD Training for Teachers and Parents Nationally
Do Good/ Earn Esteem	Esteem of Tutored Peers	Parents' Enthusiasm for LD Program	Professional Respect	Esteem of Local Schools	Satisfaction of Funder Overseers	Recognition of Scientific Contributions	Greater Sucess and Respect for Teachers
Empowerment	Leadership Status as Tutor	Prestige From Effective LD Program	National Reputation	Good Reputation With Funder	Congressional Increases in Funder Funding	Increased Prestige of Educational Psychology	Increased Power to Help LD Children
Scientific/ Clinical Success	Improved Learning Ability	Effective LD Program	Leadership Opportunities in National LD Program	Headquarters for National LD Teacher Training	Proven Success of Funded Treatment Program	Improved LD Training via Workshops	Nationally Successful LD Programs

Figure 9.1. An Example of a Benefit Matrix Based on Case 9.1: The Learning Disability (LD) Experiment

As we have seen in Chapter 8, there are many risks to be considered. Most can be minimized or avoided, and some can be transformed into benefits. The potential benefits of research are also of many kinds, including benefits to subjects, to communities, to institutions of various kinds, and, ultimately, to science and society. Just as a risk is a possible harm that one takes steps to avoid, a benefit, at the time it is discussed in the protocol, is only a potential or hoped-for good outcome that one then seeks to bring about.

This chapter examines the kinds of possible benefits, and presents a matrix of benefits and of benefit recipients; see Figure 9.1, which illustrates the matrix and specifies some of the benefits that might be developed relative to Case 9.1. Thoughtful perusal of the matrix readily suggests (a) the kinds of benefit and beneficiaries to incorporate into one's research plans, and (b) the importance of planning for specific kinds of benefit when constructing one's research proposal, rather than leaving the occurrence of benefit to happenstance. These considerations are especially important in field research where intrusion into the daily life of subjects and their community calls for reciprocity, and in risky research where commensurate benefit is required by federal law.

9.2 FEASIBLE BENEFITS AND FALSE PROMISES

Research requires much effort, and part of the spirit that carries the investigator along is the hope or belief that the research will do some good. But one should go beyond vague hope and take the necessary steps to identify and bring about as many of the potential benefits as feasible. One begins by considering the kinds of benefits that are possible and asking which of these are feasible and which can be responsibly promised in the case of one's own intended research.

9.2.1 False Promises

Benefit to society is one of the two most frequently promised benefits of research. However, most research does not lead demonstrably to an improvement in the condition of humankind. Most theses and dissertations are not even published. And most important ideas in the social sciences do not receive the follow-up required to implement them effectively in society (Sarason, 1981). Thus, one must conclude that most promises of benefit to society are based on the vague hope that all research is bound to help—somehow. Most real-world benefits to society are predicated upon more specific benefits to subjects, to the researcher, to institutions, and to science.

Benefit to science is the other most frequently promised benefit of research. To constitute such a contribution, the proposed research must address a significant problem, be based on current literature and methods, employ a valid research design, and be of such quality as to be acceptable for publication in a refereed scientific journal. (Any research that entails significant risk to subjects must meet these standards to receive IRB approval.) In short, it is unrealistic and irresponsible to promise benefit to science unless the above conditions are met.

9.2.2 Feasible Benefits

Benefit to subjects is easier to bring about than benefit to science or society. Moreover, it is the duty of researchers to give subjects something in return for their efforts. The duty to benefit subjects is particularly important in field research, wherein (a) the research represents an intrusion into subjects' daily lives, and (b) participants were given to expect some benefit when they volunteered. Much of the value of a research benefit derives from the way it is given. Thus, care, cultural sensitivity, good rapport, and good communication are vital. Important, also, is the scientific integrity of what is provided.

Scientific knowledge is a most appropriate benefit to give in return for research participation. Unfortunately, researchers often promise to give subjects the results of their study, failing to either recognize or admit (a) that the results may not be available for a long time, (b) that the results may not turn out to be very interesting or useful to subjects or to anyone else, and (c) that researchers often forget to keep such promises. Research should be based on a literature review. Hence, researchers should be able to give subjects a balanced and interesting summary of relevant knowledge at the time of their participation. That summary might take the form of a handout that is carefully edited, clear, simple, and devoid of professional jargon. The researcher should also make a cheerful and friendly offer to discuss any of the material with subjects, if they so desire. Such discussions are gratifying to subjects, and sometimes provide researchers with valuable anecdotal information that is useful in planning the next study or understanding the results of the current study. (See 4.6 for relevant discussion of debriefing.)

Personally relevant benefits are sometimes more important than scientific knowledge to subjects. Depending on the nature of the research, various kinds of personal benefits may be possible, ranging from a successful intervention that helps the subject with a personal problem, to a list of reputable local services, or an annotated bibliography that the subject would likely find useful. In the case of needy subjects, especially in community settings, personally relevant benefits might also include money, food, or medical/mental health services. Information and referrals should be offered respectfully and without implying that the subject has problems or is ignorant.

Insight, training, learning, role modeling, empowerment, and future opportunities may be natural outcomes of participation in a treatment or intervention. Opportunities for follow-up experiences or applications should be discussed with subjects. Case 1, Part III, on focus groups with crack users, exemplifies valuable benefits of this kind.

Psychosocial benefits include the benefits of (a) altruism—giving of one's time to benefit others, (b) participation in an experiment that makes one feel worthwhile, and (c) receiving favorable attention and esteem from a skilled investigator. These are typical outcomes of receiving professional and respectful treatment.

Kinship benefits include the feeling of closeness to persons, or the reduction of alienation. Research participation often relates to significant persons in the subject's life, or to people in general, and provides an opportunity to reflect on these relationships. People who are selected for participation because they have something in common—such as a learning problem, musical talent, or being a twin—may enjoy feeling that they have an opportunity, through the research, to enhance the lives of others with whom they share some human bond.

Benefits to the community or organization that is the site of field research (e.g., the subjects' school, home, neighborhood, clinic, workplace) may derive from many sources, including an actual intervention and the resources it provides, staff development, improved morale, insight into problems that need to be solved, collection of data that can be useful for policy-making or political purposes, development of new opportunities and relationships with powerful outsiders, prestige, and new abilities to serve community members. Benefits may result from a Hawthorne effect in which respect and attention paid to members of the community improve their outlook and performance. Even if a community intervention fails to produce the desired experimental effects, the project may still benefit the community in various ways.

Community-based research inevitably gets in the way of ongoing activities. The wise researcher spends considerable time learning from community gatekeepers how to accommodate to community needs and how to provide benefits expressly sought by the community—not just benefits the researcher thinks would be good for the community. The planning, and delivery, of benefits is often a condition of access to the community. Such planning is also vital to the success of an intervention, since displeased community members can sink a project; to public perception of the project; and to continuing access to that research site.

Benefits to the research institution, the researcher, and the funder are also attainable with proper planning. Equipment; ongoing training of students and qualified researchers; productive relationships with collaborators elsewhere; development of appropriate methodology; production, peer review, and publication of respected intellectual products; and public recognition of the value of the research program are needed to sustain major research programs. These kinds of benefits are not explicitly required in an IRB protocol and are rarely addressed explicitly there; and they may even be irrelevant to student projects or to isolated studies. However, IRBs understand that the researcher must build some kind of research infrastructure to sustain a research program and ultimately yield significant scientific and social benefits. Obvious neglect of the underpinnings of a research program casts doubt on promises of significant scientific and social benefit.

9.3 A MATRIX OF RESEARCH BENEFITS AND BENEFICIARIES

Significant contribution to science and society is not a one-shot activity. Rather, such contribution is typically based on a series of competently designed research or intervention efforts, which themselves are possible only because the researcher has developed appropriate institutional or community infrastructures and has disseminated the findings in a timely and effective way. Benefit to society also depends on widespread implementation, which, in turn, depends on the goodwill, support, and collective wisdom of many specific individuals, including politicians, funders, other professionals, and community leaders. Thus, the successful contributor to science and society is a builder of many benefits and a provider of those benefits to various constituencies, even if the conduct of the research per se is a solo operation.

9.3.1 The Hierarchy of Benefits

Research benefits may be divided into seven (nonexclusive) categories, ranging from those that are relatively easy to provide, through those that are extremely difficult. As shown in Figure 9.1, these seven kinds of benefits, in turn, might accrue to any of seven kinds of recipients: subjects, communities, investigators, research institutions, funders, science, or society in general. As listed below, the seven categories of benefit are described as they might pertain to a community that is the site of field research:

Valuable relationships: The community establishes ties with helping institutions and funders.

Knowledge or education: The community develops a better understanding of its own problems.

Material resources: The community makes use of research materials, equipment, and funding.

Training, employment, opportunity for advancement: Community members receive training and continue to serve as professionals or paraprofessionals within the ongoing project.

Opportunity to do good and to receive the esteem of others: The community learns how to better serve its members.

Empowerment (personal, political, and so on): The community uses findings for policy purposes; gains favorable attention of the press, politicians, and the like.

Scientific/clinical outcomes: The community provides treatment to its members (assuming that the research or intervention is successful).

Note that even if the experiment or intervention yielded disappointing results, all but the last benefit might be available to the community, as well as to individual subjects.

9.3.2 The Hierarchy of Beneficiaries

Before examining the varied forms that each of these kinds of benefits might take, let us look closer at the seven kinds of beneficiaries:

The subjects (Column 1 of Figure 9.1) are the actual research participants, whose benefits may include such things as the respect of the researcher, an interesting debriefing, money, treatment, or future opportunities for advancement. Entries in Column 1 indicate ways to enrich the lives of the subjects, engender their goodwill toward the project, and enhance their respect for science.

The community or institution (Column 2) that provides the setting for field research may include the subjects' homes, neighborhood, clinic, workplace, or recreation center. A community includes its members, gatekeepers, leaders, staff, professionals, clientele, and peers or family of the subjects. Some examples of the kinds of benefits that the community may receive have already been mentioned, both in the above example of a hierarchy of benefits (9.3.1), and in Case 9.1. Sometimes community members also serve as research assistants and so would receive benefits associated with those of the next category of recipients as well. Column 2 indicates what can be done to leave the research site a better place, satisfy community gatekeepers, and pave the way for future research.

The researcher, as well as research assistants and others who are associated with the project (Column 3), may gain valuable relationships, knowledge, expertise, access to funding, scientific recognition, and so on, if the research is competently conducted, and especially if it produces the desired result or some other dramatic outcome. By creating these benefits for oneself and other members of the research team, the investigator gains the credibility needed to go forward with a research program and to exert a significant influence upon science and society.

The research institution (Column 4) may benefit along with the researcher. Institutional benefits are more likely to be described as good university-community relations, educational leadership, funding of overhead costs, equipment, and a good scientific reputation for scientists, funders, government, and the scientific establishment. Such benefits increase a university's willingness to provide the kinds of support (e.g., space, clerical assistance, small grants, equipment, matching funds) that enable the researcher to move the research program forward with a minimum of chaos.

The funder (Column 5) is vital to the success of a major research program and hopes to receive benefits such as the following, if only the researcher remembers to provide them: ties to a good project and its constituents, well-written intellectual products promptly and effectively disseminated, good publicity, evidence of useful outcomes, good ideas to share with other scientists, and good impressions made on politicians and others who have power to reward the funder. Such benefits make a funder proud to have funded the project, eager to advertise it, and favorably disposed to funding future research of that investigator.

Science refers to the disciplines involved, and to the scientists within them, their scientific societies, and their publications. Benefits to science parallel benefits to funders and depend on the importance, rigor, and productivity of the investigation. Development of useful insights and methods may serve science, even in the absence of findings that might benefit society. Column 6 represents the kinds of events that give scientific visibility to one's ideas and empirical findings. Initial papers and symposia give way to publications and invited addresses. Others evaluate, replicate, promote, and build upon the work, thus earning it a place in the realm of scientific ideas. *A single publication upon which no one builds is not a contribution to science.*

Society, including the target population from which subjects were sampled and to which the results are to be generalized, is the one group that benefits only when the hoped-for scientific outcome occurs and is generalizable to other settings. Column 7 represents the most abstract level of benefit possible; it also reflects the most advanced developmental stage of any given research project. By the time benefits of this magnitude have accrued, the researcher or others have already adapted and broadly implemented the idea in society. The idea has begun to take on a life of its own, be modified to a variety of uses, and be adapted, used, and even claimed by many others.

The conjunction of seven kinds of benefits and seven kinds of beneficiaries yields a 49-cell matrix that is useful in research planning. This matrix suggests that turning a research idea into a scientific and social contribution requires that benefits be developed at each stage of the process. It is useful to design a tentative matrix of benefits as the basic research idea and design are being formulated and to continue planning the benefits as the project proceeds. Many valuable benefits may be easily incorporated, provided that one is attuned to opportunities for doing so.

9.4 USING THE MATRIX TO PLAN PROTOCOLS AND PROPOSALS

The ethical investigator understands the necessity of identifying the intended benefits and beneficiaries and actively striving to make those benefits a reality. Some benefits are ones that can be produced directly and easily. Others are difficult to produce, and one's best efforts to produce them may be foiled by happenstance, politics, or contrary empirical findings. Systematic planning of relevant benefit is integral to designing the research, writing the research proposal, and writing the IRB protocol. It is this specific planning process to which we now turn.

The matrix contains seven benefits, ordered approximately according to the difficulty of producing each, and seven beneficiaries, ordered approximately according to the difficulty of serving each. It provides a heuristic tool for planning to provide feasible benefits. The upper left corner of the matrix readily suggests benefits one can most responsibly promise. As one approaches the lower right corner of the matrix, one's hoped-for benefits increasingly depend upon success at prior stages of the research, on the investigator's future hard work, and on events outside the investigator's own direct control. All 49 of these kinds of benefit are valuable, though some are inapplicable to certain kinds of research. Each represents a special kind of opportunity for success; but unless these opportunities are identified and planned for, they are unlikely to materialize. Each of these kinds of benefits may be important to mention in an application for funding. Many kinds of benefits, especially those to subjects and their communities, are important to mention in the IRB protocol.

The specific benefits that one designates within the matrix depend upon the nature and magnitude of one's project. Case 9.1 provides an interesting opportunity for the reader to try out skills of identifying benefits, as this case readily yields various entries for each of the 49 cells. Figure 9.1 illustrates 49 examples of benefits that might follow from Case 9.1. The reader can no doubt identify others, and will note that certain events may pertain to more than one beneficiary and provide more than one kind of benefit.

9.4.1 The Discussion of Benefits in the IRB Protocol

The discussion of benefits in the IRB protocol should focus primarily on the benefits that are within the researcher's power and willingness to provide, especially those benefits pertaining to subjects and their community. The protocol should also allude, with appropriate modesty, to the possible scientific and social benefits that the researcher hopes will ensue from the project. This should be accompanied by a description of the methodology that demonstrates the rigor of the research and the specific anticipated products.

Perhaps the project is a major one that has developed the resources needed to undertake a ground-breaking effort that probably will influence science and social policy. Will an IRB permit the research to go forth if it necessarily involves risk? To show that the project has a distinct possibility of benefiting science and society, relevant parts of the entire matrix might be mentioned: the community goodwill and resources developed so far, the prior research and proven validity of the design and procedures, the competently trained assistants, the noteworthy prior publications, and the institutional commitment of needed resources. Thus, the IRB can see that the promise of social and scientific benefit is believable.

Or perhaps the project is a master's thesis, which is unlikely to be published. Such a project has only a slim likelihood of benefiting science and society. Therefore, the emphasis in the protocol discussion of benefit should be on what, realistically, will be done to benefit subjects, the community, and the researcher, with only modest mention of hoped-for benefits to science and society.

9.4.2 The Discussion of Benefits in the Research Proposal

Protocols should mention relevant elements from the benefit matrix: What will be done to assure benefit to subjects and the host community, to assure their ethical treatment, and to enlist their cooperation? What competencies does the investigator bring? What resources will the institution provide? What additional resources are needed? If funding is sought, what intellectual products and other benefits will the funder, science, and society receive? Who will review and evaluate these products for scientific of social merit?

PART IV

Vulnerable Populations

It is beyond the scope of this book to discuss issues and safeguards pertinent to all of the vulnerable populations mentioned in Chapter 8 (8.5). However, two vulnerable groups in particular are the topics of much research, are especially vulnerable, and are discussed here: minors and disenfranchised urban populations.

In research on youngsters, issues of risk assessment, parental permission, assent, and confidentiality are far more complex than in research with adults. These problems are discussed in Chapter 10.

Disenfranchised persons who are at risk for HIV infection include the homeless, members of the urban drug culture, poor minorities, prostitutes, runaway children, gay males, and the mentally ill. These persons pose special problems because most researchers are typically not sensitive to the culture—the beliefs, worldview, time frame, language, social structure, needs, and fears—of these people. Consider the special needs and fears that an investigator must respect in research on runaway gay male children who are homeless minorities, mentally ill, and sell their bodies for drugs. No social scientist has enough natural empathy to know how to communicate effectively with such populations and how to translate ethical principles appropriately. Chapter 11 orients the reader to ethical and procedural issues in community-based research on these populations, including ways to increase one's empathy and cultural sensitivity to such research populations.

Various other vulnerable research populations require special consideration, such as the dying, sick newborns, medically indigent populations in public hospitals, institutionalized mentally infirm, prisoners, the physically handicapped, and those in advanced stages of HIV infection. Although these populations are typically the focus of medical research, social scientists are becoming increasingly involved. Some of the same guidelines for cultural sensitivity presented in Chapter 11 apply to these populations as well. However, it is beyond the scope of this slim volume to provide the additional chapters that would be required both to present case examples and adequately describe the problems of studying these other vulnerable populations. The interested reader is referred to the following excellent sources on these and related vulnerable populations: Bermant and Wheeler (1987); Gallagher (1990); Koocher and Keith-Spiegel (1990); Levine (1986); Macklin (1987); Pope and Basque (1991); and Rothman (1991).

10

Research on Children and Adolescents[1]

As a research population, children and adolescents are special in several respects: (a) They have a limited psychological, as well as legal, capacity to give informed consent; (b) they may be cognitively, socially, and emotionally immature; (c) there are external constraints on their self-determination and independent decision making; (d) they have unequal power in relation to authorities, such as parents, teachers, and researchers; (e) parents and institutions, as well as the youngsters themselves, have an interest in their research participation; and (f) national priorities for research on children and adolescents include research on drug users, runaways, pregnant teenagers, and other sensitive topics, compounding the ethical and legal problems surrounding research on minors.

Regulations governing research respond to these characteristics of youngsters by requiring that they have special protections and that parental rights be respected. The law also expects social scientists to respond to these characteristics by using knowledge of human development to reduce risk and vulnerability.

10.1 LEGAL CONSTRAINTS ON RESEARCH ON MINORS

In 1983 the Department of Health and Human Services adopted federal regulations governing behavioral research on persons under the age of consent, 18 years.[2] These regulations include the following requirements: (a) IRB approval; (b) the documented permission of a parent or guardian, and the assent of the child–in the case of risky research or at the IRB's discretion, consent of both parents is required–and (c) that the research involve no greater risks than those ordinarily encountered in the child's daily life, except when the IRB finds that the risk is justified by anticipated benefit to the subjects (as discussed in

45 CFR 46.405). However, some research is exempted from review at the discretion of the IRB, and some of these exemptions are especially relevant to school children. Hence, a partial list of the exemptions contained in 45 CFR 46.101(b) is provided here. The reader is referred to the full set of regulations for further details. Research of the following kinds is exempted:

1. Research conducted in established or commonly accepted educational settings, involving normal educational practices such as (i) research on regular and special education instructional strategies, or (ii) research on the effectiveness of, or the comparison among, instructional techniques, curricula, or classroom management methods.

2. Research involving the use of educational tests (cognitive, diagnostic, aptitude, achievement), if information taken from these sources is recorded in such a manner that subjects cannot be identified, directly or through identifiers linked to the subjects.

Waiver of parental permission. There are two circumstances under which parental or guardian permission may be waived, at the discretion of the IRB:

1. Parental permission may be waived for research involving only minimal risk (i.e., no greater than the risks of everyday life), provided the research will not adversely affect the rights or welfare of the subjects, and provided the research could not practically be carried out without the waiver. There are various circumstances in which it is impossible or impractical to contact the parents. For example, street children who are drug dealers may or may not have parents who are available or even living; and the subjects would probably make it impossible for the researcher to make, or act on, this determination in any case.

2. Parental permission may be waived if it will not operate to protect the child. For example, abusive or neglectful parents cannot be counted on to act in their child's best interests. Parents who are in an adversarial stance vis-à-vis their child present a different problem; typically they are feeling angry or punitive because of the youngster's misbehavior. Waiver of parental permission may be appropriate under these conditions, especially when the youngster is being treated for abuse or neglect, is identified legally as incorrigible or delinquent, or is in the custody of a hospital or other institution.

Although not recognized in the federal regulations, the National Commission for the Protection of Human Subjects of Biomedical and

Behavioral Research identified four other circumstances in which waiver or modification of the parental permission might be appropriate:

1. Research designed to identify factors related to the incidence or treatment of certain conditions in adolescents for which, in certain jurisdictions, they may legally receive treatment without parental permission.
2. Research in which the subjects are mature minors, and the procedures involved entail essentially no more than minimal risk of the kind that such individuals might reasonable assume on their own.
3. Research designed to understand and meet the needs of neglected or abused children, or children designated by their parents as "in need of supervision."
4. Research involving children whose parents are legally or functionally incompetent.

The National Commission's report goes on to say that

> [T]here is no single mechanism that can be substituted for parental permission in every instance. In some cases the consent of mature minors should be sufficient. . . .
>
> In other cases, court approval may be required. Another alternative might be to appoint a social worker, nurse, or physician to act as surrogate parent when the research is designed, for example, to study neglected or battered children. Such surrogate parents would be expected to participate not only in the process of soliciting the children's cooperation, but also in the conduct of the research, in order to provide reassurance for the subjects and to intervene or support their desires to withdraw if participation should become too stressful. (43 Federal Register 2084, 1978)

In interpreting these regulations and recommendations, the Task Force for Research on Children (see Stanley & Sieber, 1991) concluded that parental incompetence is unlikely to be an adequate reason, by itself, for waiving the requirement of parental permission. Children of legally incompetent parents have guardians whose permission should be sought. When parents' deficiencies are not sufficient to result in appointment of guardians, their deficiencies probably are not grounds for abrogating their right to decide whether their children may participate in research.

The determination of "mature minor" must be made on a case-by-case basis, rather than by category. For example, a state may not rule that all 17-year-olds are mature for purposes of exercising a particular right, such as the right to abortion. Although many states grant certain rights to mature minors, all state laws concerning the rights of mature minors are silent on the subject of participation in research. Thus, there is no clear legal definition of a mature minor with respect to research participation.

Waiver of child's assent. The IRB may waive the requirement of assent if it determines that the child is incapable of assenting, due to age, level of maturity, or psychological state, or if obtaining assent would render the research impossible. However, waiver of assent is permitted only if the research involves minimal risk, or if it holds out the prospect of direct benefit to the child that is not obtainable otherwise.

Research involving greater than minimal risk. Recognizing that there may be appropriate research interventions or treatments that involve greater than minimal risk, these are permitted if they present the prospect of direct benefit to the child, if the risk is justified by the anticipated benefit, and if there are no reasonable alternatives presenting less risk. If there is no prospect of direct benefit to the subject, the research may be conducted only if (a) the intervention or procedure presents experiences to subjects that are commensurate with those of their actual or expected medical, dental, psychological, social, or educational situations, or (b) it is likely to yield vitally important generalizable knowledge about the subjects' condition.

Research involving more than minor risk beyond minimal risk, without any direct benefit, may be conducted only if approved by a panel of experts appointed by the Secretary of Health and Human Services (45 CFR 46.406).

Beyond the explicit legal provisions governing research on children, researchers are required to adhere to the general provisions governing human subject research—including consideration of risk and benefit, privacy and confidentiality, and consent. Because children are different from adults in many ways, sensitivity to developmental issues is critical; this is the topic to which we turn next.

10.2 RISK FROM A DEVELOPMENTAL PERSPECTIVE

The question to ask about vulnerability to research risk is not: At what age does a youngster cease to be especially vulnerable? Rather, it is: *How do type of risk and maturity interact?* For example, the young child, in contrast to the adolescent, is not easily embarrassed and is unlikely to be stressed by concern about the researcher's intentions, given his or her lack of capacity for self-referent thinking. He or she is unlikely even to be aware of deception, given his or her lack of suspicion of authority figures, and probably cannot effectively be dehoaxed. In contrast, the adolescent would be highly sensitive to cues that might possibly indicate the researcher's intentions. He or she would be likely to react strongly to the knowledge that he or she had been deceived, yet would be somewhat protected by his or her skepticism. He or she would require careful debriefing (see 12.7, 12.8).

Thompson (1991), drawing on Maccoby (1983), has summarized the kinds of age by vulnerability interactions that should guide risk assessment in research on minors. The reader is referred to Thompson's excellent chapter for details; the following is a summary of that discussion:

1. Younger children are more likely to experience greater behavioral and socio-emotional disorganization accompanying stress; the older child is better able to cope and more reliant on self than on caregivers. The parent's presence may buffer the young child from stress in the research setting, but may even exacerbate the older child's stress.

2. Self-conscious emotional reactions, such as shame, guilt, and pride, emerge in the preschool years, and young children have an immature understanding of these feelings. For example, young children are likely to feel guilty in negative situations for which they are not responsible. It is not until the age of 7 or 8 that children begin to restrict these feelings to appropriate circumstances.

3. Young children's trust of authority renders them especially vulnerable to coercive manipulations and to deception. With age comes understanding of individual rights and skepticism about authority.

4. Because of their limited conceptual development, younger children may benefit relatively less from feedback, dehoaxing, and debriefing.

Because of their continuing trust in authorities, they may also be less vulnerable to heightened future sensitivity to deceit in research.

5. Older children are more vulnerable to implicit cues and pressures, but they also approach the research task with more skepticism than younger children.

6. Threats to self-concept become more stressful with age. Between about 7 and 9, when children develop an integrated self-image, the evaluations of others become increasingly important (Harter, 1983). Social comparison information begins to affect self-evaluation, and ability comes to be viewed as an enduring personal quality (Nicholls, 1978). Although young children may remain optimistic in the face of negative ability attributions, older children are likely to engage in worried self-reflection and to experience lowered self-esteem.

7. As they grow older, children become increasingly sensitive to cultural and socioeconomic biases that reflect negatively on their background, family, or prior experiences.

8. As children grow older, their concern about privacy and autonomy increases. This is a major topic, which is discussed in the next section.

10.3 PRIVACY AND AUTONOMY FROM A DEVELOPMENTAL PERSPECTIVE

As we have seen in Chapter 5, privacy is a personal and idiosyncratic matter, having to do with control of the access of others to oneself. This problem is compounded in the case of children, as parents, teachers, and others also presume to have access to the child. This topic has been treated extensively by Melton (1983, 1991), and can be touched on only briefly here.

It is unclear when access to a child becomes invasive. From the time of his or her birth, caretakers have intimate physical and psychological access to the child. Gradually, the youngster takes control of his or her privacy, first through concern for privacy of possessions and space, later through concern for privacy of information. In fact, children sometimes seize control with "Keep Out" signs and locked drawers when they begin to experience the invasiveness of others. How sensitive are researchers to children's needs for privacy? Macklin (1991) offers the following test of our respect for children's privacy: Some parents install a speaker system between the baby's bedroom and the rest of the house, so they can hear the baby cry or call. This is clearly not an invasion of a 1-year-old's privacy, but what about a 5-year-old, or an 8-year-old? What are the implications for research?

A second complication arises because research on the child is often research on the family. Researchers who are accustomed to discussing (other people's) family life often fail to recognize the degree of privacy some people accord to family matters. Melton (1991) points out that even a simple task, such an essay to be written by 8-year-old on "how I spent my summer vacation," may be an implicit request for information about his or her associations and activities; it may also intrude into private family matters, as when the child spends the summer commuting between divorced parents.

Although children's privacy is easily and often invaded, privacy is clearly important even to primary school children. Indeed, children evaluate the quality of living situations by the degree of invasion of privacy and infringement of liberty present within them (e.g., Rivlin & Wolfe, 1985). Given the significance of privacy for the maintenance of self-esteem and the development of personal identity, it is extremely important to respect children's privacy by making research no more intrusive than necessary and by obtaining children's assent before entering a zone of privacy.

As we have seen in Chapter 5, privacy is intimately connected with autonomy. Children's ability to make reasoned decisions and avoid coercion with respect to research participation does not reach adult-like levels until mid-adolescence (Weithorn, 1983). However, their ability to know when someone is intolerably "in their space," or invading their privacy, exists by about the age of 6 or 7; it is appropriate, therefore, that the federal regulations of research give children veto power over their parents' permission for research participation.

Although the age of legal consent is 18, children's ability to make rational decisions is well developed before then and is not the main issue. Thus, rationality is not the main concern. At issue, rather, is the right of parents to have a say in what happens to their children. Herein lies a set of difficult issues for the researcher. As children grow older, the expression of privacy becomes an active choice; by adolescence it is a marker of independence, and the control of information becomes very important. But what of the parent? Often, the behavior being studied is rebellious and performed without the parent's knowledge. In some cases parental permission must be waived, as the youngster would never agree to having the parent informed of his or her behavior. In other cases, the behavior and the research are known to the parent, and the youngster reveals information to the researcher that the parent might wish to know about—for example, drug abuse, sexual behavior, or risk of HIV infection. Should such information be kept confidential from the parents? Do parents have a right to know? In many cases, release of such information would actually increase the risk to the youngsters. Macklin (1991) argues persuasively that a parent's desire to know does not constitute a right, and that the only ethical grounds for disclosing such information to parents is when it is only through such action that urgent help can be obtained for the youngster.

10.4 ASSENT, CONSENT, AND PARENTAL PERMISSION

Assent is defined as a child's affirmative agreement to participate in research; mere failure to object should not be construed as assent. The standard for assent is the ability to understand, to some degree, the purpose of the research and what will happen if one participates in it.

Consent requires that persons understand the consequences of their participation and be able to weigh these consequences. By tradition, the age of consent is 18, although by early adolescence most youngsters can make adult-like decisions. Some exceptions to the age of consent include full or partial emancipation, as through marriage or living apart and being self-supporting; being declared a mature minor; or being a member of the military.

Permission means the agreement of parent(s) or guardian to the participation of the child or ward in research. *Parent* means a child's biological or adoptive parent. *Guardian* means an individual who is authorized, under applicable state or local law, to consent on behalf of a child. Parental or guardian permission must fulfill the conditions of informed consent stated in Chapter 4, and documentation of the permission is required. (Exceptions are found in individual IRB policies; for example, for unobtrusive observation of school children, written permission of the school, but not the parents, may be acceptable.)

Obtaining adequate assent calls for sensitivity to maturational factors. Tymchuk (1991) suggests several ways to adapt the assent process for use with young children:

1. The level of difficulty of the information presented should be commensurate with the child's level of understanding, and his or her comprehension of the material should be assessed.
2. The format should be appropriate. A storybook or videotape format may be appropriate for young children or retarded children. Repeated presentations may be needed.
3. Training in the ability to use information to make decisions will probably yield better decision making by children.

A recommended procedure. What does one say to a child of about 5 to 12 years of age when seeking assent? Here is one possibility:

Hi, [child's name].

My name is [your name], and I am trying to learn about [describe project briefly in appropriate language].

I would like you to [describe what you would ask the child to do. Use a videotape or storybook format, if appropriate. Don't use words like "help" or "cooperate," which can imply a subtle form of coercion].

Do you want to do this? [If the child does not give clear affirmative agreement to participate, you may not continue with this child.]

Do you have any questions before we start? [Answer questions clearly.]

If you want to stop at any time, just tell me. [If the child says to stop, you must stop.]

10.5 HIGH-RISK BEHAVIOR

In consequence of the tendency of troubled youngsters to defy their parents or to run away, the law recognizes that parental consent may be waived by the IRB under certain circumstances (see 10.1). In most cases, the research is conducted within an institution, such as an HIV testing site, an abortion clinic, a youth detention center, a shelter for runaway children, or a drug treatment center. The problems of obtaining meaningful consent are manifold. These problems have been discussed extensively by Grisso (1991), who focuses on issues surrounding waiver of parental permission, and by Rotheram-Borus and Koopman (1991), who are concerned primarily with consent issues in the research and treatment of runaway, gay, and heterosexually active youth, whose relationships with their parents are often marked by secrecy, conflict, and long absences. The following is a summary of some of their main points:

1. The youngster is unlikely to believe that the research is independent of the institution or that he or she may decline to participate with impunity.
2. The youngster is unlikely to believe promises of confidentiality, especially when he or she is in trouble with his or her parents and other authorities.
3. Issues of privacy, which are normally salient for adolescents, are likely to be even more heightened for this population.
4. Maltreated youngsters are likely to experience the research as more stressful than normal children. If the researcher effectively establishes rapport, the youngster may reach out for help; the researcher must be prepared to respond helpfully.

Clearly, the researcher ignores the issues of privacy and autonomy for minors only at great peril. Both the youngster and the quality of the research will be harmed unless appropriate safeguards are employed. Adequate assent procedures, in particular, are essential. Grisso (1991) has suggested a pilot assent procedure to satisfy some of these concerns.

In the pilot assent, the researcher describes the actual research process to members of the research population in the social context where the study is to be performed. (For example, he or she obtains permission to talk to adolescent clients in an abortion clinic.) He or she then asks them a set of questions to find out if they understand what concerns such research would raise for them, and so on. In so doing, the researcher (a) refines the assent process to maximize understanding, (b) provides the IRB with documentation of adequate assent, (c) discovers possible ways to improve adolescents' assent capacities, and (d) develops the actual assent form to be employed.

In addition, Rotheram-Borus and Koopman (1991) suggest that researchers of high-risk behavior should observe the following:

1. Anticipate ethical dilemmas. Keep logs of critical incidents as an aid to formulating effective problem-solving strategies and policies.
2. Hold frequent staff meetings to discuss emerging or possible problems and to train members of the group in ethical decision making.
3. Secure assent whenever possible from the youngsters, and consent from community agency staff and parents. Parental consent should be avoided only when to seek it would jeopardize the health or well-being of the youngster.

4. Take special precautions to protect confidentiality. Whenever possible, collect data anonymously.
5. Involve the community in the design of interventions. Respect community values and perspectives; otherwise, the community can prevent successful implementation of an intervention.

Much research on high-risk behavior in youngsters is conducted in community-based settings, which themselves present special challenges; research in such settings is the topic of the next chapter.

10.6 SCHOOL SETTINGS

Schools provide a convenient entry point for researchers who want to study children. Some of the kinds of research possible through schools involve minimal risk, including noninteractive anonymous observation of public behavior, secondary analysis of data, observation of classroom behavior, testing of curriculum or teaching methods, and research based on educational testing. Often IRBs will permit such research with school permission and without parental permission. However, IRBs cannot approve research unless it complies with the Buckley Amendment.

The Family Educational Rights and Privacy Act of 1979 (the Buckley Amendment) states that: "An educational agency or institution shall obtain the written consent of the parent of a student or the eligible student [if 18 or older] before disclosing personally identifiable information from educational records of a student, other than directory information . . ." Thus, for any research that involves obtaining identifiable (as opposed to anonymous) information from student records, the investigator must obtain written permission from the parents for the specific information to be released (not blanket permission giving access to any information).

Research that involves direct intervention with school children, such as research on behavior, interviews and surveys, or introduction of special classroom activities, typically requires permission from the school district and parents, as well as the assent of each child. Research that involves risk to children, but also offers the possibility of special benefits (e.g., counseling research on abused children), might require, in addition, an IRB-appointed child advocate to monitor the consent process, and perhaps also to monitor the research. At the discretion of the IRB, the advocate might be an outside therapist specializing in the treatment of children, or a school counselor. Given the bias of researchers in favor of their own experimental treatments, the IRB might ask an independent expert to judge the likely benefit to subjects of risky research.

Example of letters requesting parental permission and child's assent appear in Appendix A.

School permission must come from the school district, not from a teacher, and must be presented to the IRB in writing, on district letterhead. Investigators should check with the district office to learn the appropriate procedure for obtaining school permission. However, schools do not have the authority to consent for children to participate in research, except as stated in the first paragraph of 10.6, above.

Avoiding coercion is especially important in school research, where peer and authority pressures are especially salient. To assure that each child's participation is truly voluntary, the researcher must implement the following objectives:

1. Minimize the coercion implicit in a request to participate from parents, teachers, or other adults.
2. Minimize peer pressure and fear of ridicule for not participating.
3. Keep any reward for participating small and not valuable.

NOTES

1. I am grateful to my fellow members of the Task Force on Research on Minors, Office for Protection from Research Risks, NIH, for contributing to my knowledge of this subject, especially to Barbara Stanley, Gary Melton, and Charles MacKay, with whom I worked most closely. This chapter draws heavily on the findings of the task force.

2. Research is also governed by state law. However, the requirements of state law are almost certain to be satisfied if parental permission is obtained. Only when a researcher wants to waive parental permission is state law likely to be more restrictive than federal regulations (Areen, 1991). Most state laws are silent on the topic of participation of minors in behavioral research, but vary in their authorization of minors to obtain various types of medical treatment without parental consent.

RECOMMENDED READINGS

Fisher, C. B., & Tryon, W. W. (Eds.). (1990). *Ethics in applied developmental psychology: Emerging issues in an emerging field.* Advances in Applied Developmental Psychology. Volume 4. Norwood, NJ: Ablex.

Keith-Spiegel, P. C., & Koocher, G. P. (1990). *Children, ethics and the law.* Lincoln: University of Nebraska Press. [This comprehensive discussion of professional issues and cases includes issues surrounding psychotherapy with children, children and the courts, and other issues that are beyond the purview of this book.]

Stanley, B., & Sieber, J. E. (Eds.). (1991). *Social research on children and adolescents: Ethical issues.* Newbury Park, CA: Sage. [This book focuses on socially sensitive research in applied settings.]

11

Community-Based Research on Vulnerable Urban Populations and AIDS

Some of the most serious social problems facing our society today arise in urban community settings among such disadvantaged populations as homeless street people, runaways, unassimilated ethnic minorities, prostitutes, intravenous drug and crack users, gay men, dual-diagnosis mentally ill, alcoholics, and the developmentally disabled. Compounding their experience of poverty, lack of education, poor mental and physical health, violence, and marginality in our culture, these populations are at high risk for HIV infection. Increasingly, researchers are seeking to help members of these groups prevent the spread of AIDS, but typically lack the cultural sensitivity required to adapt methodological and ethical principles to these settings.

The purpose of this chapter is to provide insight into ways of developing respectful and effective community-based research approaches with vulnerable urban populations. It uses, as a context, AIDS-related research, which involves ethical and methodological worst cases. This chapter focuses on disenfranchised urban populations and on the kinds of attitudes and life-styles that bring these populations into conflict with the law and complicate community-based AIDS research. Procedures are discussed for creating rapport, access, and trust within cultures that are ordinarily closed to research. This chapter emphasizes the need for cultural sensitivity, collaboration, respect, and the tailoring of research procedures to the population being studied. However, this chapter also contains lessons about community-based research that are applicable to middle-class populations.

Interventions need to be created that take into account the epidemic nature of AIDS, its 8- to 11-year latency from exposure to expression, the characteristics of the populations affected, and people's capacity to believe themselves safe. Sexually active people become vulnerable to HIV infection when their partners inject drugs or are exposed to infected individuals. Yet they are likely to ignore risk until it is too late—to believe that risky activities with one's healthy-looking partner could not be dangerous, that condoms are unacceptable, and that participation in AIDS intervention is an invasion of their privacy. Further complicating matters for researchers is the fact that the epidemic moves from one distinct population or culture to another. The specific approaches that work with one population may fail with the next.

When AIDS was first discovered, it was identified as a gay disease. Since gay males may be born into any ethnic or socioeconomic group, they include many who are educated, affluent, and hold powerful positions in society—physicians, lawyers, teachers, scientists, and other professionals. Gay men were hard hit, physically and psychologically, by the epidemic and were often brutally stigmatized by others who assumed themselves to be invulnerable to AIDS. But gay men quickly mobilized effective educational campaigns based upon scientific facts, and many members of the gay community modified their behavior so that the epidemic in this population has now reached a plateau. Gay men have also demanded a say in their treatment in medical trials:

> *Case 11.1: Community Consent.* The gay community in cities where AZT trials were scheduled did not like being guinea pigs in clinical trials that seemed to them like Russian roulette. Knowing that there were blind placebo controls, they worked as a nationwide network to develop an elaborate scheme for exchanging half doses so that each might receive possible life-saving treatment. Although this plan resulted in giving each subject an effective treatment, it foiled the toxicity-testing portion of the experimental design. In response to this research dilemma, Melton, Levine, Koocher, Rosenthal, and Thompson (1988) suggested that researchers hold forums with representatives of the subject population to plan and design the research. Medical researchers did not readily accept this idea. Consequently, key members of the gay community announced to the administration of a major medical school that they could expect no participants in AIDS research unless a regular forum were held for the gay community to negotiate the research agreements with the investigators. They expected to be in on the planning stages of the research. The prestigious investigators who met with them had never before participated in bilateral research planning with subjects. They were particularly offended when members of the gay community criticized the statistical design of their research and suggested a better design. Sessions were not very

friendly. However, the physicians involved received excellent and well-reasoned advice, and ultimately incorporated that advice into their research plans (Morin, 1990).

Such well-educated and powerful contingents were not to be found in the next populations to which the epidemic spread.

Because AIDS was mistakenly considered a gay disease, the next populations to which the epidemic spread—the inner-city poor, especially blacks, Latinos, and Asians who are intravenous drug users—had no idea that they were at risk; nor were these victims acknowledged or treated with compassion by members of their own group. In the course of this second epidemic, AIDS spread to women and, since blood is exchanged between mother and fetus, to their children. Now, as the AIDS epidemic sweeps through populations of disenfranchised and uneducated people in urban areas, it begins to enter another population that is even less willing to acknowledge its vulnerability and mortality—sexually active teenagers of all ethnic and socioeconomic groups.

11.1 COMMUNITY-BASED INTERVENTION RESEARCH

The community-based scientist works collaboratively within the existing social structure of the community, focusing on prevention (rather than cure), competence building, social support, empowerment, and mutual help. Empowerment means enabling persons and groups to solve their problems and meet their needs in their own cultural context; community researchers serve and build constituencies among community members (Gesten & Jason, 1987).

The methods of community research are often experimental and behavioral and may involve types of statistical design and analysis unfamiliar to social scientists. These include time-series analyses (Steiner & Mark, 1985), social-impact assessment and nonequivalent group designs (Meissen & Cipriani, 1984), and diverse qualitative methods (e.g., Susskind & Klein, 1985).

11.1.1 Settings for Community-Based AIDS Research

The researcher typically begins with a particular intervention research objective and target population in mind. For example, the researcher may want to teach safe sex practices to prostitutes, to develop a social support system for persons seeking to stay off drugs, or to provide a needle exchange program for drug-addicted homeless people. Community-based AIDS and drug-abuse research typically occurs on the street, in the living situation of the subject, in a drug-treatment or health-care setting, or under the auspices of some existing community organization.

Street-based research means working with people such as the homeless, drug users, and runaway children to create an intervention that promises to improve their lives. The cultures and informal gatekeepers of the streets are difficult for the typical researcher to get to know. Street people distrust outsiders, even researchers of the same ethnic background who speak the same language and are formerly of the same group (e.g., former prostitutes or drug addicts). An important survival skill of street people is to take on whatever persona is needed to get what they want. The middle-class researcher is often taken in by life stories that are simply untrue. Because of the difficulties of becoming an insider to street culture, researchers often turn to community organizations for access to communities.

Community organizations, such as churches or schools, may welcome the resources of a social scientist who is willing to work within their agenda. Such organizations may have considerable power to facilitate community interventions. Thus, if the researcher becomes a trusted and integral part of the organization and empowers people within their own culture, it may be possible to create an effective and lasting intervention. The costs to the researcher may include placing members of the community organization on the project payroll, doing community service work, and assisting the community organization in obtaining its own funding. A danger of working within community organizations is that they may want to modify results in order to look good in the eyes of funders, or to base interventions on beliefs or values that blame the victim (Ryan, 1971). Some community churches, for example, believe that AIDS represents the wrath of God, and will not work compassionately with AIDS victims or persons at risk for AIDS.

Clinic-based research offers a more scientifically oriented setting for research and interventions on hospitalized or outpatient subjects. Community-based physicians and drug-abuse treatment programs offer an excellent opportunity for the researcher to work with persons who have AIDS or are at risk for AIDS, and who have a trusting relationship with the staff of a clinic that provides their health care. Like indigenous community organizations, however, clinics have their own priorities and programs. Clinics will not permit research that might reduce the level of trust between patients and clinic staff, interfere with ongoing clinic activities, violate clinic rules, or create bad public relations.

11.1.2 Community Gatekeepers

All community settings have gatekeepers. These are persons who can help the researcher to learn the community culture and enter into effective working relationships with community members—or who can keep the researcher out. A gatekeeper is a leader with the power to decide whether an outsider has the potential to bring benefit to the community and should be admitted, though an unscrupulous gatekeeper may allow research that serves the gatekeeper's needs at a cost to the community. There are also informal gatekeepers, people who have the power to sink a project if they disagree with the formal gatekeeper's decision. Thus, the wise researcher tries to try to work out cooperative arrangements that suit all of the formal and informal gatekeepers, and seeks another community site if that effort proves unsuccessful.

Gatekeepers may be scientists, such as a researcher who also directs a clinic; they may be street professionals, such as a recovered drug addict who serves as an outreach person for his own people; or they may be nonscientist professionals, such as a school principal or a minister. The gatekeeper introduces the researcher to the community culture and negotiates the terms under which the researcher may fit in.

11.1.3 Community Cultures

The most serious error a neophyte community researcher can make is to assume that street people, unemployed drug addicts, homeless runaways, and other disenfranchised members of society are necessarily unintelligent, or are out there alone. Such people live within a culture that has well-defined rules of conduct, communication, and attitudes toward outsiders. Members of the community look out for one another. They have a grapevine that rapidly spreads news about outsiders, as well as about one another; hence, clear communication and concern for the privacy of research participants are vital to the success of any program. Although many street people have unorthodox life-styles, they may be highly intelligent.

Communities of disenfranchised persons tend to distrust researchers and other professionals. Even runaway children who may be in imminent danger of being murdered tend to survive on bravado, believing that it is safer to sell their bodies for food or drugs than to trust a stranger who claims to be a helping professional. It takes a thoughtfully negotiated relationship with a gatekeeper to become trusted and accepted, as discussed in 11.4.

In clinic-based research researchers must learn to obey staff rules, get along with staff members individually, and contribute to the achievement of clinic goals. They must also understand the needs, attitudes, and behavior of the clients who might be, for example, street-wise, unruly, paranoid, manipulative, or frightened. Researchers must also respect and contribute to the clients' relationship of trust and rapport with clinic staff, which the clinic has worked hard to build, and must do nothing that would damage that trust, such as failing to respect clients' privacy or employing procedures that the clients might misunderstand or find objectionable.

To intervene and empower community members, one needs not only to get *access*, but also to *intervene effectively*. How can Latina women be enabled to persuade Latino men to use condoms? How can gay Asian men be enabled to form support groups that encourage safe sex? How can teenagers be effectively educated about birth control and safe sex? How can anyone be educated to understand that healthy-looking persons may carry the HIV virus? The simple rational educational approaches that readily occur to middle-class social scientists have often proved to be ineffective. A knowledge of the literature on intervention and cultural sensitivity (e.g., Bowser, 1990; Marin & Marin, 1991; Peterson & Marin, 1988), coupled with personal ethnographic efforts in the target community, are essential to successful intervention. The literature on AIDS and culturally sensitive intervention is growing so rapidly that perhaps the best advice to give here is to consult the most recent *Psychological Abstracts* and *Sociological Abstracts* and conduct computerized searches in the social/behavioral and medical science literatures.

11.2 WHAT IS CULTURAL SENSITIVITY?

In research on vulnerable populations, cultural sensitivity has almost nothing to do with the art and music of a culture, and almost everything to do with respect, shared decision making, and effective communication. Too often, researchers ignore the values, the life-style, and the cognitive and affective world of the subjects. They impose their own, perhaps in an attempt to reform people whose culture they would like to eradicate, or perhaps simply out of ignorance about the subjects' reality. This chapter provides various approaches to gaining cultural sensitivity. But first, what are the major sensitivities the researcher needs to gain?

Any communication or intervention must be couched in terms of the subjects' basic assumptions (not the researcher's) about such things as health, illness, AIDS, sexuality, personal adequacy and self-worth, sin and salvation, science, whose advice to take about medical or sexual matters, masculinity, femininity, and any other topic that the research touches upon. In addition, the researcher must have open lines of communication through which not only to learn community members' current views about the researcher's motives, the risks or benefits of participation, and so on, but also effectively address misconceptions. The needs and fears of the target population must be both understood and alleviated to the extend possible by the project. The social, religious, political, economic, and psychological barriers to communicating about AIDS, sexuality, illness, or other sensitive topics addressed by the research must be taken into account, and ways must be found to overcome these barriers. Knowing the subjects' time frame is vital; promises about "next year" mean nothing to people who live in the present. Knowing their family structure and dynamics are essential to knowing whose advice and authority are influential.

Concerns about control, autonomy, and exploitation will be raised by interventions that attempt to influence sexual or other life-style characteristics. It is essential to build trust, multilateral and shared decision making, and an equal-status relationship between intervener and target population. The project members who come in contact with the research participants must be acceptable to the community.

There is much potential for miscommunication when talking about sexuality, disease, and other personal topics where private or idiosyncratic vocabularies tend to abound; persons may be embarrassed to ask for clarification or to reveal their own beliefs. Ways must be found to discover and use the terminology of the target populations. Where English is not spoken fluently, the communication must be in the language of the target population.

11.3 WHY IS AIDS-RELATED RESEARCH PROBLEMATIC?

Many problematic aspects of AIDS-related research have already been illustrated. Other problems include informed consent, especially with youngsters; the illegality of the activities of many of the populations at risk for AIDS; the stigma of AIDS; problems surrounding AIDS testing; sampling problems; and the difficulty of directly observing the behaviors conducive to AIDS.

Informed consent. The requirement of parental permission for research participation (discussed in Chapter 10) may be waived by the IRB when parents are unavailable or unlikely to act in their child's best interests, or when parental consent cannot feasibly be obtained and the research promises significant benefit at little risk (Rotheram-Borus & Koopman, 1991). Waiver of parental permission is often necessary since many of the populations at risk for AIDS are youngsters who are runaways, or whose sexual orientation, sexual activity, or drug use are not known to their parents. Some of these youngsters are simply engaging in a temporary (even if unprecedentedly risky) rebellious phase designed to separate themselves from their parents. For others, the problems are deeper, involving dysfunctional families, mental illness, or mental retardation. These matters, of course, complicate intervention as well as consent. However, the problems of intervention among adolescents are beyond the scope of this chapter; the reader is referred to Woodruff, Doherty, and Athey (1989).

Other consent issues involve problems of language and culture. Any consent statement must be administered in a style entirely understandable to subjects, taking into account their perception of the risks and the benefits they wish to receive for their participation. It must be in the subjects' primary language and communicated in a way that they understand and accept. The standard consent procedure may be replaced by a consent *forum* such as a focus group, community consent, or so-called hot dog ethnography, which are described below in section 11.4. In these situations, the standard consent form is largely a formality to satisfy federal law, as the consent agreement has already been discussed and decided by the community.

The illegal activities of some populations at risk for AIDS raise additional problems concerning consent, payment of subjects, police harassment, and the psychology of fugitive populations. When community research populations are engaged in illegal behavior (e.g., prostitution, drug use) or are members of stigmatized groups (e.g., gay men, people with AIDS), it may be unreasonable to obtain signed consent. A signed consent form on file may represent a potential threat to the well-being of the subject or, more likely, it may have been signed with a pseudonym. If signatures are obtained, a certificate of confidentiality should be obtained, and the signed consent forms should be kept in a secure place, such as the investigator's safe deposit box.

Since drug addicts may be willing to do anything for money, the issue of subtle coercion must be addressed. It is widely recognized that if some form of payment is not offered, drug addicts will not participate in research. Food is an important alternative form of payment to addicts and other street populations. Although most members of street populations neglect nutrition, they typically are ravenously hungry when presented with a good meal, and eating forms an important bonding ritual between them and the research team. However, this presents a second problem for researchers: Most funders do not allow food as a line item on a budget since it normally would not be used in this way. Hence some new terminology, such as "in-kind payment," usually must be used in budgeting for food.

Police harassment of subjects and researchers is a distinct possibility. For example, at an early point in a program of prostitute education, San Francisco police seized the condoms given to prostitutes, photocopied them to use as evidence of prostitution, then punched holes in them before returning them (Lockett, 1990). Researchers should gain city support for their projects whenever possible, for example, via the mayor's office or other powerful city agencies such as the city health department. In Lockett's case, above, a trip to the mayor's office stopped police seizure of condoms.

The psychology of fugitive populations involves endless paranoia and subterfuge. For this reason, only the most experienced researchers may be able to work with their target population without the aid of gatekeepers who understand that population's perceptions, fears, and strategies.

The stigma of AIDS and other sexually transmitted diseases means that many AIDS-related projects must state their purpose in euphemistic terms—for example, The Janesville Health Study, rather than The Janesville AIDS or Syphilis Study. Communities may reject members thought to have AIDS. People fear that AIDS testing could result in notification of their sexual partners or a government agency, quarantine, arrest, curtailment of insurance and employment, or other harm. People fear that they will be found to have a sexually transmitted disease and that other members of the community will learn about it. Consequently, the privacy of participants in AIDS-related research must be protected in every conceivable way, and every effort must be made to let subjects know that they will not be stigmatized by their participation. Even the location and architecture of the building where the program is housed should be planned to shield subjects' participation in the research program from the scrutiny of community members and others who might gossip.

AIDS testing itself is surrounded by problems. Many persons do not want to know their test results. When giving test results, it is essential, and required by federal law governing human research, that results be accompanied by adequate counseling. The topics of AIDS testing and counseling are beyond the scope of this chapter; the reader is referred to Fawzy, Namir, Wolcott, Mitsuyasu, and Gottlieb (1989); Meinhart (1989); Morin (1990); and Pope and Morin (1990).

Sampling problems in some populations at risk for AIDS are unresolvable. There are no adequate sampling frames for intravenous drug users, prostitutes, runaway youngsters, and so on. In some cases, the populations are transient, moving from one part of a city to another as the availability of drugs or services changes, so that even a geographically based sampling scheme would be inadequate. In other cases, these populations are hidden (Bowser, 1990). Such populations include intravenous drug users who are successful in the straight world and do not see themselves as drug addicts; closeted Latino and Asian homosexuals who hide their orientation from their community; and men who view themselves as heterosexuals, but engage in occasional sex with men. These hidden populations are difficult to reach. They are not clearly defined and may not define themselves as risk-takers. Their behavior (were it known to others) would be negatively viewed by the general public. Hence, they are underground in a private world.

The most rigorous sampling methods are not workable with some populations at risk for AIDS. Samples of subjects are often gathered through community recruitment; by posting an ad on a community bulletin board; or by being where the action is, for example, at a needle exchange site, at a place where condoms and food are handed out, or at a lunch and rap session. Even after obtaining volunteers, however, investigators often need to maintain careful observation to ascertain that they indeed meet the sampling criteria. A screening interview, for which all volunteers are paid, might be used to try to identify those who falsified some characteristic simply to be paid (e.g., gays who said they were straight, sexually active girls who claimed to be virgins).

Behaviors associated with AIDS, for example, injection drug use and sexual behavior, cannot acceptably be studied by unobtrusive methods and typically are not behaviors that persons discuss candidly. Clinic-based research may be essential for determining persons' actual status or behavior. For example, youngsters who go to a clinic for family planning, venereal disease, or drug-abuse may participate candidly in research, but in other contexts may be unwilling to reveal their sexual activity or drug abuse. The National Institute of Allergies and Infectious Diseases has published a useful paper suggesting ways to combine behavioral and epidemiological strategies for intervention research on these behaviors.

11.4 TECHNIQUES FOR CREATING CONSENT AND SHARED TRUST

Collaboration and shared trust are essential to community-based interventions. A consent *forum*, rather than a consent form, is required. Three basic kinds of forums are described here. In 11.5, these ideas are further developed in descriptions of relationships with various kinds of community gatekeepers.

11.4.1 Community Consultation and Consent

Effective community research requires consultation with the target population. The basic idea is that accepted representatives are chosen by their community to negotiate the conditions of the research in an open forum with the researcher. The objectives and concerns of the researcher, subjects, and target population are examined. Goals, methods, and procedures are revised until they are acceptable to both sides. Few traditionally trained researchers expect to take advice from subjects on matters of research design and procedure.

However, the rules are different in community-based research, and often the quality of research and subject cooperation are improved with community consultation. Community consultation is discussed extensively by Melton et al. (1988), and an example of subject-initiated community consent was described in Case 11.1 (the gay community's negotiations with a medical school). Consent forums with less educated populations are described in 11.4.2 and 11.4.3.

11.4.2 Focus Groups

Focus group techniques are useful for learning the views and concerns of one's target population. The use of focus groups to achieve the purposes of community consent have already been discussed in Case 1, Part III (Bowser's use of focus groups with teen crack users). Focus groups, especially centered around a meal, are an outstanding way for a researcher to learn about a population and its community, and to develop collaborative research or intervention plans. In the process, trust and rapport are developed. The plans that are formulated in focus group meetings are sure to be disseminated throughout the community via the grapevine, resulting in further feedback from the community. Focus group techniques are discussed more extensively in Morgan (1988) and Stewart and Shamdasani (1990).

11.4.3 "Hot Dog Ethnography"

A simple consent forum consists of giving a picnic for one's target population and inviting its collaboration in a project:

> *Case 11.2: Hot Dog Ethnography.* A researcher/intervener wanted to establish a needle exchange program among street people and needed to know about her clientele and their concerns: Are they diverse populations or are they homogeneous? How will she identify individuals for repeated measures (i.e., whether they continue to exchange needles)? How can she best arrange to meet with them and avoid police arrest for exchanging needles? What are their needs and concerns? Which of these can she satisfy?
>
> There was no established gatekeeper except some street people who were respected by their peers. (The researcher subsequently served as gatekeeper for others who became involved in the needle

exchange program.) After some conversations with individuals in the area where she considered establishing the needle exchange, she issued a word-of-mouth invitation to dinner. She rented a hotel room in a flophouse and cooked hot dogs, sauerkraut, and soup; she also served cake, donuts, and soda pop in abundance. Her hot dog ethnography, as it has been called, was a great success. Most street people are quite hungry. About 40 people attended, which is surprising, given that they are a stigmatized population and police entrapment was a possibility. The researcher was known to them from her previous outreach efforts, which helped engender trust. The researcher learned that some of her clientele lived in flophouses and preferred to come to a designated place for the weekly needle exchange. Others lived on the streets and wanted a floating exchange place where the police would be less likely to wait for them. The willingness of the researcher to risk arrest was testimony to her caring for the population she sought to serve. A successful needle exchange program was established and has lasted for 4 years (Case, 1990).

11.5 WORKING WITH GATEKEEPERS

We turn now to specific examples of work with gatekeepers.

11.5.1 Street-Based Research

Locating a street-based gatekeeper who will welcome one's particular approach to intervention and research is partly a matter of being referred to the appropriate gatekeeper and partly a matter of negotiation. There might be a referral from a social agency, a community intervener, a researcher, or the street people themselves. The negotiation is a matter of learning whether the match between gatekeeper and researcher can be made into a compatible one. Each negotiation is different, but the following case is illustrative:

Case 11.3: Food and Condoms. A researcher wanted to learn what prostitutes understand about safe sex, whether it is possible to get them to use condoms, how many are seropositive, and whether drug-addicted prostitutes differ in their sexual behavior from career prostitutes. She established a working relationship with the director of a prostitute education and advocacy organization. The director, a black ex-prostitute, employs the following procedure to serve her clientele. She meets with groups of prostitutes and (a) serves them food, (b) asks them what she may do to help them, (c) finds out particular problems they may be experiencing, (d) works out ways to solve the problems, (e) assesses kinds of education they may need and enjoyable ways to provide that education (e.g., quizzes on safe sex, with nice prizes for those who get all the answers right), and (f) establishes regular places for prostitutes to meet, eat, voice concerns, and obtain needed resources. But what about the researcher? She made her research approach fit within this context of helping and educating, worked as a faithful team member with the gatekeeper, and achieved her objectives of survey research, HIV testing, and AIDS education (Lockett, 1990).

11.5.2 Clinic-Based Research

Clinics and their gatekeepers have many serious concerns the researcher must satisfy. Among these are concern for maintaining confidentiality of treatment files, managing their potentially unruly clients, keeping the researcher from interfering with the duties of the staff, making sure the clients are treated respectfully, and ensuring that nothing is done that casts doubt on the loyalty of the clinic to its clients. In addition, most clinics receive some federal funding and are prohibited by the government from permitting certain activities on their premises; for example, needle exchange research is prohibited at the time of this writing. Sorensen (1991) suggests a four-step approach to working with clinic-based gatekeepers:

1. Approach the clinic's research coordinator before proposing the study for funding, or IRB approval. Explain the research idea and its potential benefits to patients, the clinic, and science, as well as risks or inconveniences it may entail. Learn whether the idea is at all acceptable and the conditions under which the clinic is likely to participate.

2. Obtain a letter of contingent approval—a letter granting approval provided that certain conditions are met (e.g., time, space, clinic management and staff approval, funding, IRB approval).

3. Obtain consent from the administrative group that conveys actual clinic approval of research. At this stage, details will be worked out,

such as researcher access to patient charts, approval from a local research committee, number and kind of subjects to be made available to the researcher, and so on. If these negotiations are successful, obtain a letter of intent to cooperate, stating not only what each party will give and receive through the study but also any further issues to be negotiated. This letter should accompany any request for funding or IRB approval, and should be signed by both the clinic leader and the researcher.

4. Begin meeting with the staff to resolve such issues as who may participate, how much time the clinic will allot for each interview, and when and where interviews will occur. Unless the researcher is attentive to the needs of the staff, their passive resistance will destroy the study. The wise researcher attends all clinic staff meetings and works to make the research an integral service of the clinic's program, rather than an extra burden for staff and patients. Careful attention to Columns 1 and 2 of the benefit matrix (Figure 9.1) is amply repaid.

11.5.3 Other Agencies and Institutions

Although other agencies and institutions may have fewer rules than clinics, that may only be because they are unaccustomed to accommodating researchers. The lack of formal rules may result in misunderstanding and a premature end to the research program. It is better to try to go through steps such as those described for clinic-based research and to get the agreement in writing than to leave matters to chance. In nonscientific settings, it is especially important to reach a signed agreement on who has the right to decide on the formal content of presentations of the findings, and who has access to confidential data.

11.6 RESEARCH DESIGN

Some design considerations follow from the characteristics of the vulnerable populations described here (Sieber & Sorensen, 1992). Many of the theories and methods of traditional social science are unacceptable, including research that blames the victim (Ryan, 1971), deception strategies, procedures that may raise suspicions, studies requiring much reading or writing, research that provides no benefit for subjects, and some uses of standard psychological tests (Huang, Watters,

& Case, 1988). Extensive pilot testing is always advisable, and the researcher should build into the timetable provision for various redesigns of the study, including ones necessitated after the formal research has begun. Community-based research is full of surprises that threaten the validity of the research.

Long-term follow-up and random assignment may be difficult to employ. Drug users are difficult to locate for repeated measures unless one works out a strategy for doing so during the initial testing. For example, subjects may be asked for information on how to reach someone who will always know where they are. Subjects should be promised sufficient payment for the follow-up to motivate them to return when contacted.

Random assignment should not deprive anyone of desired benefits, and should be explained clearly. Otherwise, subjects will suspect deception and may engage in subterfuge to obtain the sought-after benefit.

11.7 WHAT TO DISCUSS IN THE PROTOCOL

In addition to the usual contents of a protocol, the IRB will need to know how access to the research population was obtained and what agreements were made with gatekeepers. They will want to be sure that the researcher has adequate knowledge of the target population; has done some pilot testing; and understands the actual risks to subjects, their fears of risk, and risks to the researcher.

The protocol should describe whatever kind of consent forum has been used, for example, focus groups, hot dog ethnography, meetings with gatekeepers. Written consent of gatekeepers should be presented, where appropriate. The actual consent form should reflect the agreements that were worked out in the consent forum.

Confidentiality is always a concern in community research, and every step that has been taken to ensure confidentiality should be described, including the location and configuration of the research site, the training of research assistants, agreements with gatekeepers, and the safekeeping of data. Where repeated measures are involved, special attention should be given to ways of maintaining anonymity or confidentiality while matching pre- and post-test data (see Chapter 6). Street people, especially those on drugs, are likely to forget which pseudonym they used on the pretest, or to lose the sheet of paper containing their code number.

Researchers using repeated measures sometimes use informal methods for matching pre- and post-test data, such as a private written description of individuals in conjunction with their names or code numbers. One does not want to mix up data, and in particular one does not want to mix up the results of HIV tests.

Special arrangements, such as waiver of signed consent by subjects, waiver of parental permission, or the obtaining of a certificate of confidentiality, may be necessary and should be explained and justified in full.

The IRB may be unfamiliar with the research population, the complexities of community-based research, and the gatekeepers who control the setting. Thus, for example, the researcher may need to educate the IRB about such things as the fact that research procedures may need to be changed in response to new problems, and that gatekeepers may later raise new issues not reflected in the original protocol. It is not unusual, in community-based research, for a researcher to need to submit a preliminary protocol, followed by modifications. The researcher who does a conscientious job of submitting the initial protocol will likely have the full cooperation and sympathy of the IRB when modifications need to be approved later.

RECOMMENDED READINGS

Edwards, J., Tindale, R. S., Heath, L., & Posavac, E. J. (Eds.). (1992). *Social psychological applications to social issues: Vol. 2. Methodological issues in applied social psychology.* New York: Plenum.

Herdt, G., & Lindenbaum, S. (Eds.). (1991). *Social analysis in the time of AIDS.* Newbury Park, CA: Sage.

Marin, G., & Marin, B. (1991). *Research with Hispanic populations.* Newbury Park, CA: Sage.

PART V

Developing an Effective Human Subjects Protocol

The protocol demonstrates to the IRB, and to anyone who might inquire, that the research is respectful of the needs and interests of the subjects, that risks have been reduced to a minimum, and that the benefits of the research more than justify any risk, inconvenience, or other cost it might create. The key to developing an effective protocol is to incorporate relevant ethical considerations into one's early planning of the project, and to begin writing the protocol along with the research proposal. Chapter 12 provides guidelines for developing a protocol, but cannot substitute for careful incorporation of the ideas from Chapters 3 through 11 into the research plan.

12

Developing a Research Protocol

The protocol format and reminders offered in this chapter combine the best of many of the forms used around the nation and will be useful even if the reader's own institution has a specific form. An IRB may raise any questions pertinent to the specific research project, even if those questions are not specifically addressed in their particular protocol format or instructions. This chapter, and its references to the rest of the book, enable the reader to answer such questions.

12.1 SUGGESTED ELEMENTS OF A PROTOCOL

One's individual protocol should reflect the requirements of one's department and IRB and should contain any additional information pertinent to the evaluation of the particular project. The following protocol elements meet federal requirements and include additional features that many institutions have found useful.

1. A *cover sheet* should include (a) the name and department of the principal investigator (PI), (b) his or her faculty rank or student status, (c) home and office phone number(s) and (d) address(es) of the investigator. This information enables the IRB reviewer to contact the investigator informally about questions that arise when reading the protocol, and perhaps to provide verbal approval before the formal approval is mailed. The cover sheet should also indicate (e) the project title; (f) the type of project, such as faculty research, externally funded project (with name of funder), student directed research (with name of faculty adviser, thesis, dissertation, course requirement—give course number and faculty name); and (g) the intended project starting and ending dates. It is useful to mention one's qualifications to conduct the specified research in a paragraph or two on the cover sheet, via an attached curriculum vitae, or with the description of the methodology. The cover sheet must contain (h) the signature of the principal investigator, and if the PI is a student, (i) the adviser. Some institutions also designate departmental or school representatives to review and sign off on protocols.

2. A *description of the research*, which includes the following: (a) the purpose of the research and the hypotheses to be tested; (b) the historical background of the research, referring to pertinent scientific literature (in brief, as in an abstract); (c) an orderly account of the research method, design, and mode of analysis, detailed enough that reviewers can assess scientific validity, including a fully detailed account of procedures that directly affect subjects; (d) a realistic statement of the value of the research, including both what the researcher expects to learn from the research and what value it will have for the participants and their community, the research institution, the funder, or science (Research is not of value to science unless it is of publishable quality. See Chapter 9.); (e) the location of the research—specifying the exact laboratory, community, institution, and so on where various components of the research are to be performed, the reason why that setting was chosen and how the researcher happens to have access to it; (Is the researcher employed there? Did he or she do volunteer work there? Is it his or her old neighborhood or an organization to which he or she belongs?; (f) duration of the project and how this window of time coincides with such other time constraints as the duration of funding, the periods of the school year when research can reasonably be carried out in a school, the period of time before an election when voting attitudes might be examined.

3. A *description of the prospective subject population* should include, where relevant, ethnic background, sex, age, and state of health. It should explain why that particular population is being used, the source(s) from which it will be obtained, and a statement of the selection criteria. If vulnerable populations are included, such as pregnant women, children, institutionalized mentally disabled, prisoners, or those whose ability to give voluntary informed consent is in question, the rationale for using such subjects should be stated. If the research is conducted in an institutional setting (e.g., a school, a club, a church, a home for the aged), written permission of the person in charge must accompany the protocol.

The expected number of subjects should be specified, and a statistical justification of the number of subjects should be provided either here or in the description of the research design. The researcher is urged to consult Kraemer and Thiemann (1987) and Lipsey (1990) for guidelines to deciding how many subjects to use.

4. The *discussion of possible risks* should include inconveniences or discomforts, especially to the subject, and where possible, an estimate

of the likelihood and magnitude of harm. Most IRB members are highly skilled risk assessors and take a dim view of researchers who ignore minor risks or inconveniences and blithely write "no risk." There are many forms of risk to subjects and others connected with the research, including the investigator, the community, and the institution. These are discussed in Chapters 5 (Privacy), 6 (Confidentiality), 7 (Deception), 8 (Categories of Risk), 10 (Research on Children), and 11 (Community-Based Research).

Discussion of risks should involve both objective risks and what subjects might perceive as risks (as in Case 1.1), and should indicate what will be done to allay each actual risk or unwarranted worry. As appropriate, the researcher might describe alternative methods that could have been used to minimize risk, stating why they were rejected. For example, IRBs are always quick to urge that data be collected anonymously to prevent breach of confidentiality; however, the researcher may have good reasons to collect unique identifiers.

5. *Discussion of inducements and benefits* to the subject and others should take into account the concepts regarding benefit, presented in Chapter 9, and field research, in Chapter 11.

6. *Freedom of subjects to withdraw with impunity* is a right that must, by law, be respected. If the subject is not free to withdraw from the research at any time, the protocol should both explain why and state when the subject is free to withdraw. Pertinent details of subjects' freedom to withdraw should appear in the consent statement.

7. *Source and amount of compensation*, if any, to be received by a subject or beneficiary in the event of injury is typically not addressed in social research protocols, where chances of injury are very small and liability for incidental injury is often covered by the university's workmen's compensation insurance.

8. *Analyses of risks and benefits* are to be summarized, and any risk must be shown to be substantially offset by benefits that the researcher has arranged to produce.

9. The *informed consent* procedure should be described, including how, where, and by whom informed consent will be negotiated, and how debriefing will be conducted (see Chapter 4). In Chapter 10, problems of obtaining children's assent and parental or guardian permission for research are discussed.

The actual consent form, if any, should be attached to the protocol. If consent is negotiated orally and not documented in writing, a statement should be attached regarding the information that is to be presented to prospective subjects orally. The content of the debriefing should also be described.

10. *Attachments*—such as any letters of permission, the consent form, interview or survey questions, materials to be presented to subjects, tests, or other items connected with the research—should be attached to the protocol, if they might be pertinent to the IRB's evaluation of the project.

12.2 SUPPLEMENTAL PROTOCOL CHECKLIST

The following is a list of typical problems that IRBs encounter, and tips on how to cure them:

"Rubber stamp" signatures. Those who sign off on the cover sheet of the protocol have a legal responsibility to have evaluated the protocol for legality, clarity, accuracy, and good writing. Students dissatisfied with their formal supervision should seek other help with the protocol and the design and analysis of the research, and consult research methods texts as needed.

Failure to mention investigator qualifications. This is especially serious in the case of inexperienced investigators. This statement should be clear, specific, and relevant. It might include prior research training and experience, membership in or special knowledge of the research population, or qualifications for counseling subjects as appropriate.

Too many generalities about the purpose of the research. IRBs are suspicious of protocols that devote much space to extolling the importance of the research, but fail to describe the methods and procedures adequately.

Vagueness about research location and permission. The protocol should be specific about where the research will be performed and how the investigator got permission to do the research. It should include letter(s) of permission from the relevant gatekeepers and discussion of what the researcher has agreed to do in return for that permission.

Vagueness about sampling procedure. Arrangements for access to subjects and for sampling should be complete before submitting the protocol. The protocol should be exact in the description of the sample frame, how it was obtained, and exactly what sampling procedure will be employed (e.g., a two-phase random sample using a table of random numbers, a random sample stratified on ethnic group with oversampling of Native Americans, a convenience sample); see, for example, Babbie

(1979) or Kidder (1981). If the sample frame consists of members of a private group, the IRB will want to know if the list of names is public information, and if not, how the researcher obtained that list and what permission was obtained to use it. Depending on the sensitivity of the research, some sampling strategies may pose objectionable threats to privacy; see Hartley (1982) for details.

Vagueness about the research design. The protocol should state exactly what general design or method is to be employed; for unusual procedures, it should describe any work previously done to test the procedure. Complex research designs might be accompanied with diagrams, if necessary, to show who gets what treatment and when, and when measurements are taken. For survey research, specify who and how many are to be surveyed, when and by what method they are to be surveyed, and what key cross-tabulations are planned. For case studies, state whether it is a behavioral single-subject design or a clinical case study; indicate how the raw data will be obtained and how the case study will be derived from those data, perhaps citing a methodology text that sets forth the rules. Protocols for action research should state why the action research is called for, the specific goals, the activities that are expected to achieve those goals, and how those goals are expected to come about. The goals of the action research should be ones jointly developed with the subjects, not goals foisted on subjects.

Omission of information about the political context of the research. Often, applied social research is done because there are conflicts, problems, or disagreements between parties at the research site. Perhaps the research is done to understand something about the problem, or to intervene. It is essential that the political context of the research be described accurately in the protocol.

"Fitting the format." Most protocol formats are designed for experimental or descriptive research, not for action or intervention research. In order to "fit the protocol," researchers sometimes make their project look like it is an experimental or descriptive study when it is actually action or intervention research. The protocol should state clearly what the purpose of the research is, even if the protocol format seems designed for describing something else.

Ignoring risk. Most social research involves some risk, if only that a survey may ask people to think about things that will make them uncomfortable, or that the data on some subjects, were it to fall into the hands of a malicious gossip, could cause trouble for the subjects. IRBs recognize that some risk is inevitable and acceptable; what they find unacceptable is the researcher who fails to recognize risk.

Insensitivity to issues of coercion in dual-role relationships. It is often easiest to arrange to do research in familiar settings; however, this is likely to involve a dual-role relationship. For example, one might study one's own clients, students, or employees or do participant observation research in a group of which one is a part. Or one may arrange to do research in a group where the gatekeeper takes unscrupulous advantage of the situation (perhaps coercing members of his or her organization to participate, or seeking access to confidential data in return for permitting the researcher access to the setting).

IRBs recognize that the only feasible way to do some important kinds of applied research is in a dual-role relationship. For example, a graduate student cannot easily arrange to try out a teaching intervention on someone else's class, or to try out a therapy intervention on someone else's patients. In such situations, however, special precautions must be taken:

- Every step must be taken to assure that subjects know that their participation is strictly voluntary—that they will lose no advantage and will fully retain the respect and goodwill of the researcher and gatekeeper if they refuse to participate.
- Subjects must have a neutral source to whom they can turn in case of problems. For example, if free counseling is offered to anyone who is upset by his or her research participation, the counseling must be available from an independent third party, not the researcher or the gatekeeper.
- Where feasible, participation should be anonymous, so that the researcher or gatekeeper does not know who participated or who did not.

A second kind of dual role is that in which the researcher is also an intervener. The researcher needs to be clear about what he or she and others consider to be his or her primary role. If the researcher is primarily an intervener—one who provides a service—then relevant other services may not be withheld from subjects.

Dual-role relationships introduce potential conflicts, which should be recognized at the outset and discussed not only with experienced members of one's IRB but also with experienced researchers who have had to work in dual-role situations.

Using data generated by others. When some or all of one's data have been generated by others, the IRB will want to know both the source of the data (e.g., a public archive, an individual scientist, a school or university student testing program) and who released the data and authorized their use in the proposed research. A letter of authorization may be required.

Research on physical and physiological qualities. Research on the effects of, say, caffeine or physical exertion may be safe for most, but not all, subjects. The IRB will want to know that the campus physician or some other medically qualified individual has reviewed and approved the research plan.

Research that stigmatizes persons. The researcher who is intent on helping persons who are in need of some intervention is likely to overlook the fact that research participation may stigmatize the subjects. So-called prevention research, community interventions, behavior modification programs, and research on people who already occupy a status to which stigma is attached may heighten the visibility of the these people's stigma. Every effort must be made to ensure the privacy of such individuals.

Appendix: Sample Consent and Assent Forms for Use With Older Children

The following sample letters were adapted from ones developed by the IRB at SUNY Albany:

PARENTAL PERMISSION FORM

[School Letterhead]

Date:_______

Student's Name: ______________________ Grade:________

Dear Parent:

Our school is participating in a study conducted by Jane Jones, a graduate student at Western University. The project is titled: Verbal Processing in Elementary School Readers. The study compares children reading below grade level with those reading at or above grade level on various measures of learning and memory.

Your child has been selected based on the testing to which you previously agreed.

With your permission, your child will work with a person from the Study Center on six occasions, for approximately 20 to 30 minutes each time.

During each session, your child will be presented with a variety of tasks designed to measure attention, memory, language, learning, visual-spatial, and motor ability. The tasks are not difficult and in most instances the children find them quite enjoyable.

Each child will be seen on a one-to-one basis, and scheduling will be arranged with the teacher to make sure your child does not miss important classroom activities. Performance on all of the tasks will be kept confidential.

If you have any questions regarding the study, please contact Ms. Smith, Research Coordinator, at 555-1111.

Below is a form for you to sign. Please indicate whether you agree to have your child participate, and have your child return the form to school tomorrow.

Your cooperation in this research would be greatly appreciated.
Sincerely,

Principal

I ____ give
____ do not give permission to have my child, ____________________
(child's name)
participate in the study involving verbal memory. I understand the nature of the study and the amount of time involved.

STUDENT ASSENT

Student's Name:__________________________ School:______________

[The researcher sits down with each child separately, hands the child the following written material, and reads it aloud as the child follows along.]

Do you remember the permission slip you took home for your parents to sign a few days ago?

The people I work with and I are interested in how people learn about words. We are asking you and other kids to help us find out about it.

If you agree, I will need you to help me six times: today, and five more times over the next few weeks. We will work together for about 20 to 30 minutes each time. We will be doing something different each time, but you shouldn't have any difficulty with any of the things we do. Sometimes we'll ask you to find certain letters or numbers on a page or to remember letters of words that you see or hear. Other times, I'll ask you to draw some figures, follow directions from a map, tell some stories, and listen to some sentences to tell me what they mean.

This is not a test like you usually have in school. All you have to do is try as hard as you can to do the things I ask and you'll do fine. Your teachers and parents and the other children will not know how well you do. It will be just between you and me and the people I work with.

I would really appreciate it if you would help me to find out about these things, but, if for some reason you feel like you really don't want to do this, just tell me. You may quit at any time.

Do you have any questions? [The researcher should answer any questions the child asks, without discussing specific test items.]

If you agree to work on these reading tasks with me, I would like you to sign this paper. It says: [Researcher reads assent statement.]

Date:_________

The information above has been read to me and any questions I had have been answered. I would like to take part in the activities that have just been described to me.

____
Student's Signature

[Note: A signed Assent Form is not always appropriate and depends on the age of the child and the nature of the research. Consult your IRB about obtaining children's signatures.]

References

American Psychological Association. (1973). *Ethical principles in the conduct of research with human participants.* Washington, DC: Author.

American Psychological Association. (1982). *Ethical principles in the conduct of research with human participants.* Washington, DC: Author.

Areen, J. (1991). Legal constraints on research on children. In B. Stanley & J. E. Sieber (Eds.), *The ethics of research on children and adolescents.* Newbury Park, CA: Sage.

Asch, S. (1956). Studies of independence and conformity: A minority of one against a unanimous majority. *Psychological Monographs, 76*(9), Whole 416.

Babbie, E. R. (1979). *The practice of social research.* Belmont, CA: Wadsworth.

Bermant, G., & Wheeler, R. R. (1987). From within the system: Educational and research programs at the Federal Judicial Center. In G. B. Melton (Ed.), *Reforming the law* (pp. 102-145). New York: Guilford Press.

Boruch, R. F. (1976). *Methodological techniques for assuring personal integrity in social research.* Unpublished manuscript (NIE-C63).

Boruch, R. F., & Cecil, J. S. (1979). *Assuring the confidentiality of social research data.* Philadelphia: University of Pennsylvania Press.

Boruch, R. F., & Cecil, J. S. (1982). Statistical strategies for preserving privacy in direct inquiry. In J. E. Sieber (Ed.), *The ethics of social research: Surveys and experiments.* New York: Springer-Verlag.

Bowser, B. P. (1990). AIDS and "hidden" populations. *MIRA: Multicultural Inquiry and Research on AIDS, 4*(1), 1-2.

Campbell, D. T., Boruch, R. F., Schwartz, R. D., & Steinberg, J. (1977). Confidentiality-preserving modes of access to files and to interfile exchange for useful statistical analysis. *Evaluation Quarterly, 1*(2), 269-300.

Case, P. (1990). The prevention point needle exchange program. In J. E. Sieber, Y. Song-Kim, & P. Kelzer (Eds.), *Vulnerable populations and AIDS: Ethical and procedural requirements for social and behavioral research and intervention.* Hayward, CA: Pioneer Bookstore.

Conner, R. F. (1982). Random assignment of clients in social experimentation. In J. E. Sieber (Ed.), *The ethics of social research: Surveys and experiments* (pp. 55-77). New York: Springer-Verlag.

Cox, L. H., & Boruch, R. F. (1986). Emerging policy issues in record linkage and privacy. In *Proceedings of the international statistical institute* (pp. 9.2.1-9.2.16). Amsterdam:ISI.

Davies, S. P. (1930). *Social control of the mentally deficient.* New York: Thomas Y. Crowell.

Diener, E., & Crandall, R. (1978). *Ethics in social and behavioral research.* Chicago: University of Chicago Press.

Doris, J. (1982). Social science and advocacy: A case study. In J. Sieber (Ed.), Values and applied social science [Special Issue]. *American Behavioral Scientist, 26,* 199-233.

Duncan, G., & Lambert, D. (1987, April). The risk of disclosure for microdata. *Proceedings of the Third Annual Research Conference*. U. S. Bureau of the Census.

Fawzy, F. I., Namir, S., Wolcott, D. L., Mitsuyasu, R. T., & Gottlieb, M. S. (1989). The relationship between medical and psychological status in newly diagnosed gay men with AIDS. *Psychiatric Medicine, 7*, 23-33.

Fernald, W. E. (1919). A state program for the care of the mentally defective. *Mental Hygiene, 3*, 566-574.

Fienberg, S., Martin, M., & Straf, M. (1985). *Sharing research data.* Washington, DC: National Academy Press.

Fisher, C. B., & Rosendahl, S. A. (1990). Psychological risks and remedies of research participation. In C. G. Fisher & W. W. Tryon (Eds.), *Ethics in applied developmental psychology: Emerging issues in an emerging field. Advances in Applied Developmental Psychology. Volume 4* (pp. 43-59). Norwood, NJ: Ablex.

Fox, J. A., & Tracy, P. E. (1986). *Randomized response: A method for sensitive surveys.* Beverly Hills, CA: Sage.

Gallagher, H. G. (1990). *By trust betrayed: Patient-physician and the license to kill in the Third Reich.* New York: Holt, Rinehart & Winston.

Gates, G. W. (1988, August). *Census Bureau microdata: Providing useful research data while protecting the anonymity of respondents.* Paper presented at the annual meeting of the American Statistical Association, New Orleans.

Geller, D. (1978). Involvement in role-playing simulations: A demonstration with studies on obedience. *Journal of Personality and Social Psychology, 36*, 219-235.

Geller, D. (1982). Alternatives to deception: Why, what and how? In Sieber, J. E. (Ed.), *The ethics of social research: Surveys and experiments* (pp. 38-55). New York: Springer-Verlag.

Gesten, E. L., & Jason, L. A. (1987). Social and community interventions. *Annual Review of Psychology, 38*, 427-460.

Goldman, P., Clark, E., & Marro, A. (1975, July 21). No one told them: Suicide of F. R. Olson linked to CIA drug experiment. *Time*, pp. 15-19.

Grant, W. V., & Eiden, L. J. (1980). *Digest of Education Statistics, 26*, 199-233.

Greenawalt, K. (1974). Privacy and its legal protections. *Hastings Center Studies, 2*, 45-68.

Grisso, T. (1991). Minors' assent to behavioral research without parental permission. In B. Stanley & J. E. Sieber (Eds.), *The ethics of research on children and adolescents* (pp. 109-127). Newbury Park, CA: Sage.

Harter, S. (1983). Developmental perspectives on the self-system. In P. H. Mussen (Ed.) (E. M. Hetherington, Vol. Ed.), *Handbook of child psychology, Vol IV. Socialization, personality, and social development* (pp. 275-385). New York: John Wiley.

Hartley, S. F. (1982). Sampling strategies and the threat to privacy. In J. E. Sieber (Ed.), *The ethics of social research: Surveys and experiments* (pp. 167-189). New York: Springer-Verlag.

Heller, J. (1972, July 26). Syphilis victims in US study without therapy for 40 years. *The New York Times*, pp. 1, 8.

Holmes, D. (1976). Debriefing after psychological experiments: Effectiveness of post experimental desensitizing. *American Psychologist, 32*, 868-875.

Hook, E. G. (1973). Behavioral implications of the human XYY genotype. *Science, 179*, 139-150.

Huang, K.H.C., Watters, J., & Case, P. (1988). Psychological assessment and AIDS research with intravenous drug users: Challenges in measurement. *Journal of Psychoactive Drugs, 20*, 191-195.

Humphreys, L. (1970). *Tearoom trade: Impersonal sex in public places.* Chicago: Aldine.

Isen, A., & Levin, P. (1972). Effect of feeling good on helping: Cookies and kindness. *Journal of Personality and Social Psychology, 21*, 384-388.

Jones, J. (1982). *Bad blood.* New York: Free Press.

Katz, J. (1972). *Experimentation with human beings.* New York: Russell Sage.

Kelman, H. (1967). Human use of human subjects: The problem of deception in social psychological experiments. *Psychological Bulletin, 67*, 1-11.

Kelman, H. (1972). The rights of the subject in social research: An analysis in terms of relative power and legitimacy. *American Psychologist, 27*, 989-1016.

Kelman, H. C. (1968). *A time to speak: On human values and social research.* San Francisco: Jossey-Bass.

Kevles, D. J. (1970). Into hostile political camps: The reorganization of international science in World War I. *Isis, 62*, 47-60.

Kidder, L. H. (1981). *Research methods in social relations.* New York: Holt, Rinehart & Winston.

Kim, J. (1986). A method for limiting disclosure in microdata based on random noise and transformation. *1986 Proceedings of the Survey Methodology Research Section* (pp. 370-374). Washington, DC: American Statistical Association.

Klockars, C. B. (1974). *The professional fence.* New York: Free Press.

Koocher, G. P., & Keith-Speigel, P. C. (1990). *Children, ethics and the law: Professional issues and cases.* Lincoln: University of Nebraska Press.

Kraemer, H. C., & Thiemann, S. (1987). *How many subjects? Statistical power analysis in research.* Newbury Park, CA: Sage.

Laufer, R. S., & Wolfe, M. (1977). Privacy as a concept and a social issue: A multidimensional developmental theory. *Journal of Social Issues, 33*, 44-87.

Levine, R. J. (1986). *Ethics and regulation of clinical research.* Baltimore, MD: Urban & Schwarzenberg.

Lipsey, M. (1990). *Design sensitivity: Statistical power for experimental research.* Newbury Park, CA: Sage.

Lockett, G. (1990). AIDS prevention with Cal-PEP, COYOTE, and Project Aware. In J. E. Sieber, Y. Song-Kim, & P. Kelzer (Eds.), *Vulnerable populations and AIDS: Ethical and procedural requirements for social and behavioral research and intervetions.* Hayward, CA: Pioneer Bookstore.

Loo, C. M. (1982). Vulnerable populations: Case studies in crowding research. In J. E. Sieber (Ed.), *The ethics of social research: Surveys and experiments.* New York: Springer-Verlag.

Maccoby, E. E. (1983). Social-emotional development and response to stressors. In N. Garmezy & M. Rutter (Eds.), *Stress, coping and development in children* (pp. 217-234). New York: McGraw-Hill.

Macklin, R. (1987). *Mortal choices: Bioethics in today's world.* New York: Pantheon.

Macklin, R. (1991). Autonomy, beneficence and child development: An ethical analysis. In B. Stanley & J. E. Sieber (Eds.), *The ethics of research on children and adolescents.* Newbury Park, CA: Sage.

Marin, G., & Marin, B. V. (1991). *Research with Hispanic populations.* Newbury Park, CA: Sage.

McKusick, L., Wiley, J., & Coates, T. J. (1985). AIDS and the sexual behavior reported by gay men in San Francisco. *American Journal of Public Health, 75*, 493-496.

Meinhart, R. A. (1989, November). AIDS and issues of partner notification. *FOCUS: A Guide to AIDS Research and Counseling*, 1-2.

Meissen, G. J., & Cipriani, J. A. (1984). Community psychology and social impact assessment: An action model. *Journal of Community Psychology, 12*, 369-386.

Melton, G. B. (1983). Minors and privacy: Are legal and psychological concepts compatible? *Nebraska Law Review, 62*, 455-493.

Melton, G. B. (1990). Brief research report: Certificate of confidentiality under the Public Health Service Act: Strong protection but not enough. *Violence and Victims, 5*(1), 67-70.

Melton, G. B. (1991). Respecting boundaries: Minors, privacy and behavioral research. In B. Stanley & J. E. Sieber (Eds.), *The ethics of research on children and adolescents.* Newbury Park, CA: Sage.

Melton, G. B., Levine, R. J., Koocher, G. P., Rosenthal, R., & Thompson, W. C. (1988). Community consultation in socially sensitive research: Lessons from clinical trials on treatments for AIDS. *American Psychologist, 43*, 573-581.

Melton, G. B., & Stanley, B. H. (1991). Research involving special populations. In B. H. Stanley, J. E. Sieber, & G. B. Melton (Eds), *Psychology and research ethics.* Lincoln: University of Nebraska Press.

Mirvis, P. H. (1982). Know thyself and what thou art doing. *American Behavioral Scientist, 26*(2), 177-197.

Mitroff, I. I., & Kilmann, R. H. (1979). *Methodological approaches to social science.* San Francisco: Jossey-Bass.

Morgan, D. L. (1988). *Focus groups as qualitative research.* Newbury Park, CA: Sage.

Morin, S. (1990). Behavioral research on gay men in San Francisco. In J. E. Sieber, Y. Song-Kim, & P. Kelzer (Eds.), *Vulnerable populations and AIDS: Ethical and procedural requirements for social and behavioral research and intervention.* Hayward, CA: Pioneer Bookstore.

National Commission for Protection of Human Subjects of Biomedical and Behavioral Research. (1978). *The Belmont Report: Ethical principles and guidelines for the protection of human subjects of research* (DHEW Publication No. (OS) 78-0012). Washington, DC: Government Printing Office.

Nicholls, J. G. (1978). The development of the concepts of effort and ability, perception of academic attainment, and the understanding that difficult tasks require more ability. *Child Development, 49*, 800-814.

Orne, M. (1969). Demand characteristics and the concept of quasi-controls. In R. Rosenthal & R. Rosnow (Eds.), *Artifacts in behavioral research.* New York: Academic Press.

Pelto, P. J. (1988, February 18-20). In J. E. Sieber (Ed.), *Proceedings of a conference on sharing social research data.* National Science Foundation/American Association for the Advancement of Science, Washington, DC. Unpublished.

Peterson, J., & Marin, G. (1988). Issues in the prevention of AIDS among black and Hispanic men. *American Psychologist, 43*, 871-877.

Pope, K. S., & Basque, M.J.T. (1991). *Ethics of psychotherapy and counseling: A practical guide for psychologists.* San Francisco: Jossey Bass.

Pope, K. S., & Morin, S. F. (1990). AIDS and HIV infection update: New research, ethical responsibilities, evolving legal frameworks, and published resources. *The Independent Practitioner, 10*, 43-53.

Rivlin, L. G., & Wolfe, M. (1985). *Institutional settings in children's lives*. New York: John Wiley.

Rosenthal, R., & Rosnow, R. L. (1969). *Artifact and behavioral research*. New York: Academic Press.

Rotheram-Borus, M. J., & Koopman, C. (1991). Protecting children's rights in AIDS research. In B. Stanley & J. E. Sieber (Eds.), *The ethics of research on children and adolescents*. Newbury Park, CA: Sage.

Rothman, D. J. (1991). *Strangers at the bedside: A history of how law and bioethics transformed medical decision making*. New York: Basic Books.

Ryan, W. (1971). *Blaming the victim*. New York: Pantheon.

Sarason, S. B. (1981). *Psychology misdirected*. New York: Free Press.

Sarason, S. B., & Doris, J. (1969). *Psychological problems in mental deficiency*. New York: Harper & Row.

Sarason, S. B., & Doris, J. (1979). *Educational handicap, public policy, and social history*. New York: Free Press.

Seeman, J. (1969). Deception in psychological research. *American Psychologist, 24*, 1025-1028.

Sieber, J. E. (1989). Sharing scientific data. I: New problems for IRBs to solve. *IRB: A Review of Human Subjects Research, 11*(6), 4-7.

Sieber, J. E., & Saks, M. J. (1989). A census of subject pool characteristics and policies. *American Psychologist, 44*(7), 1051-1063.

Sieber, J. E., & Sorensen, J. L. (1992). Ethical issues in community-based research and intervention. In J. Edwards, R. S. Tindale, L. Heath, & E. J. Posavac (Eds.), *Social psychology applications to social issues. Vol. 2: Methodological issues in applied social psychology*. New York: Plenum.

Sorensen, J. L. (1991). Gatekeeping AIDS research in drug treatment programs. In J. E. Sieber, Y. Song-Kim, & P. Kelzer (Eds.), *Vulnerable populations and AIDS: Ethical and procedural requirements for social and behavioral research and intervention*. Hayward, CA: Pioneer Bookstore.

Stanley, B., & Sieber, J. E. (1991). Epilogue. In B. Stanley & J. E. Sieber (Eds.), *The ethics of research on children and adolescents*. Newbury Park, CA: Sage.

Stanley, B. H., & Guido, J. R. (1991). Informed consent: Psychological and empirical issues. In B. H. Stanley, J. E. Sieber, & G. B. Melton (Eds.), *Psychology and research ethics*. Lincoln: University of Nebraska Press.

Steiner, D. D., & Mark, M. M. (1985). The impact of a community action group: An illustration of the potential of time series analysis of for the study of community groups. *American Journal of Community Psychology, 13*, 13-30.

Stewart, D. W., & Shamdasani, P. N. (1990). *Focus groups: Theory and practice*. Newbury Park, CA: Sage.

Susskind, E. C., & Klein, D. C. (Eds.), (1985). *Community research: Methods, paradigms, and applications*. New York: Praeger.

Thompson, R. A. (1991). Vulnerability in research; A developmental perspective on research risk. In B. Stanley & J. E. Sieber (Eds.), *The ethics of research on children and adolescents*. Newbury Park, CA: Sage.

Tobach, E., Gianutsos, J., Topoff, H. R., & Gross, C. G. (1974). *The four horsemen: Racism, sexism, militarism and social Darwinism.* New York: Behavioral Publications.

Turner, A. G. (1982). What subjects of survey research believe about confidentiality. In J. E. Sieber (Ed.), *The ethics of social research: Surveys and experiments* (pp. 151-166). New York: Springer-Verlag.

Tymchuk, A. J. (1991). Assent processes. In B. Stanley & J. E. Sieber (Eds.), *Social research on children and adolescents: Ethical issues.* Newbury Park, CA: Sage.

Vinacke, W. (1954). Deceiving experimental subjects. *American Psychologist, 9,* 155.

von Hoffman, N. (1970, January 30). Sociological snoopers. *Washington Post*, pp. B1, B9.

Weithorn, L. (1983). Children's capacities to decide about participation in research. *IRB: A Review of Human Subjects Research, 5,* 1-5.

White, D. (1991). Sharing anthropological data with peers and third world host countries. In J. E. Sieber (Ed.), *Sharing social science data: Advantages and challenges.* Newbury Park, CA: Sage.

Wiener, S., & Sutherland, G. (1968). A normal XYY man. *Lancet*, 1359.

Wolosin, R., Sherman, S., & Mynatt, C. (1972). Perceived social influence in a conformity situation. *Journal of Personality and Social Psychology, 23,* 184-191.

Woodruff, J. O., Doherty, D., & Athey, J. G. (1989). *Troubled adolescents and HIV infection: Issues in prevention and treatment.* Washington, DC: CASSP Technical Assistance Center, Georgetown University Child Development Center.

Zimbardo, P., Haney, C., Banks, W., & Jaffe, D. (1973, April 8). The mind is a formidable jailer: A Pirandellian prison. *The New York Times Magazine*, pp. 38-60.

Author Index

Subject Index

About the Author

Joan E. Sieber is Professor of Psychology at California State University, Hayward, and the recipient of that university's Outstanding Professor Award for 1991. A social psychologist by training, her area of specialization is the study of emerging ethical and value problems in social research and intervention. Her main interest is in the development of procedural and methodological solutions to ethical problems in human research that otherwise would limit the value of social science. In addition to her work on ethical issues in human research, she currently works on problems of developing culturally sensitive approaches for outreach, research, and intervention with vulnerable populations at risk for AIDS, problems of research on powerful people, and problems of promoting data sharing among scientists. She is the author of the Sage volume *Sharing Social Science Data: Advantages and Challenges*, and co-editor with Barbara Stanley of the Sage volume *Social Research on Children and Adolescents: Ethical Issues.*

Printed in the United States
58531LVS00002B/31

9 780803 939646